SpringerBriefs in Computer Science

Series editors

Stan Zdonik

Shashi Shekhar

Jonathan Katz

Xindong Wu

Lakhmi C. Jain

David Padua

Xuemin (Sherman) Shen

Borko Furht

V.S. Subrahmanian

Martial Hebert

Katsushi Ikeuchi

Bruno Siciliano

Sushil Jajodia

Newton Lee

More information about this series at http://www.springer.com/series/10028

Joel Gibson · Oge Marques

Optical Flow and Trajectory Estimation Methods

 Springer

Joel Gibson
Blackmagic Design
Colorado Springs, CO
USA

Oge Marques
Department of Computer and Electrical
  Engineering
Florida Atlantic University
Boca Raton, FL
USA

ISSN 2191-5768   ISSN 2191-5776   (electronic)
SpringerBriefs in Computer Science
ISBN 978-3-319-44940-1   ISBN 978-3-319-44941-8   (eBook)
DOI 10.1007/978-3-319-44941-8

Library of Congress Control Number: 2016948603

Printed on acid-free paper

This Springer imprint is published by Springer Nature
The registered company is Springer International Publishing AG
The registered company address is: Gewerbestrasse 11, 6330 Cham, Switzerland

To Andrea

—Joel Gibson

To Ingrid

—Oge Marques

Preface

Optical flow can be thought of as the projection of 3-D motion onto a 2-D image plane. We are generally given the 2-D projections of the objects at different points in time, i.e., the images, and asked to ascertain the motion between these projections. While points of a physical object, considered at different points in time, should indeed have some dense motion vector in 3-D space, the projection of these points onto a 2-D image sacrifices this one-to-one characteristic. Indeed, it is ordinary that the projection of a point on an object is hidden or *occluded* from view or moved outside of the domain of the image. This inverse process is akin to trying to deduce objects from shadows cast on the ground.

Yet understanding the motion within a scene is the key to solving many problems. Within film and video, high-quality motion estimation is fundamental to the restoration of archival footage. Optical flow is used to help interpolate frames for speed changes. Robots use real-time motion approximations to navigate their environment. Combined with stereo depth estimation, optical flow is an intrinsic part of scene flow.

Perhaps the most fundamental concept in optical flow is *color constancy*. It claims that the projection of a given point on any image will produce the same color value. For all but synthetically generated images this will not hold exactly. As the amount or angle of light changes between the captured images, color intensity can vary dramatically. A closely related property which attempts to mitigate this variability is *gradient constancy* or edge matching.

For all but the most simplistic case, matching the color of a pixel between two images yields a one-to-many map. The process might be improved by comparing the neighborhoods around a pixel in order to find a best match. This too fails along the edge of an object where the neighborhoods all look the same in what is called *aperture effect.*

We must add some knowledge about how flow behaves in order to choose which of the many possible color constancy matches is best. In this role, the most successful regularizer has been Total Variation (TV). Roughly speaking, total variation in optical flow sums the total amount of change in the flow field. Then, given ambiguous flows with similar color constancy, it will choose the flow with the least total change.

Since the beginning of modern optical flow estimation methods, multiple frames have been used in an effort to improve the computation of motion. More recently, researchers have stitched together sequences of optical flow fields to create *trajectories*. These trajectories are temporally coherent, a necessary property for virtually every real-world application of optical flow. New methods compute these trajectories directly using variational methods and low-rank constraints.

Optical flow and trajectories are ill-posed, under-constrained, inverse problems. Sparse regularization has enjoyed some success with other problems in computer vision but there has been little application in optical flow. In part this is because of the difficulty of dictionary learning in the absence of an exemplar. Applying sparsity to trajectories as a low-rank constraint has been stifled by the computational complexity.

This book focuses on two main problems in the domain of optical flow and trajectory estimation: (i) The problem of finding convex optimization methods to apply sparsity to optical flow; and (ii) The problem of how to extend sparsity to improve trajectories in a computationally tractable way.

It is targeted at researchers and practitioners in the fields of engineering and computer science. It caters particularly to new researchers looking for cutting-edge topics in optical flow as well as veterans of optical flow wishing to learn of the latest advances in multi-frame methods.

We expect that the book will fulfill its goal of serving as a preliminary reference on the subject. Readers who want to deepen their understanding of specific topics will find more than eighty references to additional sources of related information spread throughout the book.

We would like to thank Courtney Dramis and her staff at Springer for their support throughout this project.

Colorado Springs, CO, USA   Joel Gibson
Boca Raton, FL, USA   Oge Marques
June 2016

Contents

Chapter 1
Optical Flow Fundamentals

Abstract In this chapter we cover concepts that appear repeatedly in algorithms in the remainder of this book. Most of the topics are native to 2-frame optical flow methods but are also relevant to multi-frame flow and trajectories. In 2-frame methods the relative motion is computed from a *reference* frame to a *target* frame. The target does not have to be the next frame after the reference frame, indeed, it could precede the reference frame.

1.1 Color Constancy

Color constancy is the quite reasonable expectation that a point in a reference frame will have the same color at its new location in a target frame. Color constancy alone is quite unreliable. There are typically many locations in a target frame that will closely match a brightness (or color) of a reference pixel. Indeed, the best match is not necessarily the correct one. Brox and Malik [2] point out that a nearest neighbor algorithm will satisfy the color constancy constraint admirably but will produce completely useless flow vectors. If a point is not occluded in the target frame, we expect the color to approximately generally match the reference point. However, this modest goal often fails due to defects such as lighting changes and specular highlights. We must rely on some form of regularization to choose flow vectors that exhibit spatial or temporal smoothness while simultaneously balancing the color constancy constraint.

Mathematically color constancy for a single brightness is

$$I(x, y, t) = I(x + u, y + v, t + 1) \tag{1.1}$$

where $I(x, y, t)$ is the intensity of an image for a given location in space and time. Here u, v represent a horizontal and vertical motion again at a specific pixel in space and time although for simplicity we do not write $u(x, y, t)$ or $v(x, y, t)$.

The gradient of the image is often used to supplement color constancy. Edges are less sensitive to lighting changes. Sometimes a structure/texture separation [1] is performed as a preprocessing step. The two are then blended back together with a

© The Author(s) 2016

J. Gibson and O. Marques, *Optical Flow and Trajectory Estimation Methods*,
SpringerBriefs in Computer Science, DOI 10.1007/978-3-319-44941-8_1

strong emphasis on texture. This captures some of the benefits of color and gradient constancy with less computation in the optimization stage.

Most algorithms use a single brightness channel for color constancy. Using multiple color channels does show some improvement [3, 7], but somewhat surprisingly, the improvement is small. Color channels are highly correlated and the color constancy defects previously mentioned are likely to be present in all the channels.

Horn and Schunck's original work [4] used a quadratic penalty on color constancy. This was likely chosen for its tractability at the time, but this over-penalizes outliers. Color constancy errors are not Gaussian, i.e. small variances in the image or flow produce corresponding small changes in the color constancy error. The L^1 or linear penalty norm is more tolerant of outliers and is often called a robust penalty. Upgrading the color constancy to an L^1 norm, which is non-smooth, complicates the optimization, but gives dramatically better results.

Figure 1.1b shows an example of color constancy only informing us of the magnitude flow perpendicular to an image gradient. The magnitude of flow along image gradient is unknown. The homogeneous regions must be filled using some assumption about the flow. One popular choice assumes flow is generally smooth, except at edges in the image.

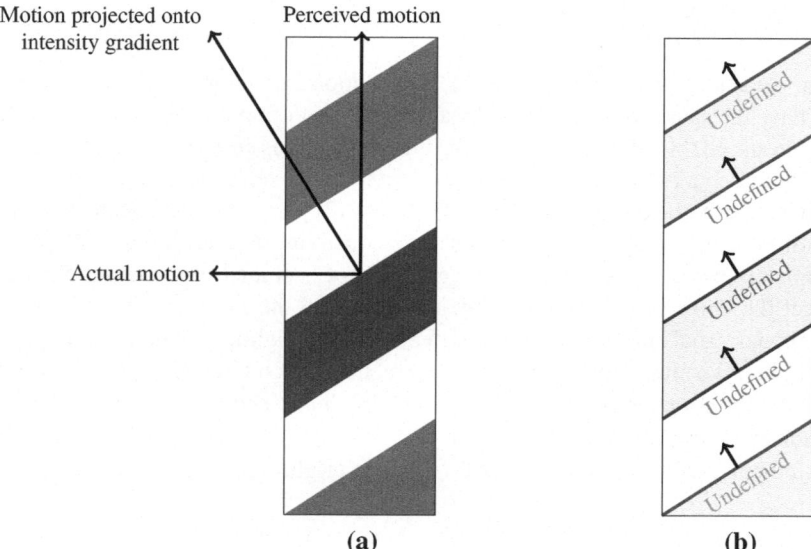

(a)   (b)

Fig. 1.1 Aperature effect and color constancy shortcomings. The aperature effect in **a** makes any of the proposed directions viable, **b** illustrates that even with a flow direction color constancy only yields flow at the edges requiring some kind of smoothing to fill the undefined majority

1.2 Aperture Problem

The aperture problem refers to the ambiguity that results from viewing a homogeneous edge of a larger object through a smaller view port or aperture. If the edge is observed to be moving, it is possible to know the magnitude of the motion perpendicular to the edge but it is not possible to know the magnitude of the motion perpendicular to the edge. Figure 1.1a shows a typical example of of aperture effect. For additional illustrations see Palmer [6].

The beginning of the aperture problem can be seen in Eq. 1.1 which is underconstrained with two variables for each equation.

1.3 Small Versus Large Motion

The color constancy constraint is highly nonconvex, i.e. the color constancy error does not get larger and larger the farther away from the correct flow vector one strays. Indeed it may get much better near a false match. The choices made to address this challenge strongly effect the properties of optical flow algorithm.

1.3.1 Linearization

A common approach to the color constancy constraint is to take a first-order Taylor series approximation of the color constancy constraint.

$$u\frac{\partial I}{\partial x} + v\frac{\partial I}{\partial y} + \frac{\partial I}{\partial t} = 0. \tag{1.2}$$

This formulation is convex and amenable to computationally efficient solutions. However, this approximation is only valid for small motions of less than one pixel. To make this useful, a coarse-to-fine image pyramid is made with processing starting on the coarse scale. Now, a small relative motion can correspond to a large motion in the original image. The flow vectors are propagated to the next finer stage. Additionally, the target image is *warped* by the intermediate flow result such that the vectors that are being solved for are the small difference between the warped image and the reference image.

This coarse-to-fine approach breaks when there is large motion by small objects. In the linearized model, this large motion can only be captured while at the coarse scale. If the object or structure is too small to be visible at the coarse scale then its motion will not be captured and the finer scales will have no chance of recovering the correct displacement.

1.3.2   Nonconvexity

Another approach is to avoid the linearization of the color constancy constraint. The usual stategy is to solve (approximately) the Euler-Lagrange equations of the objective function. Since the L^1 norm is not differentiable, it is replaced with a differentiable approximation, for example, by $\Psi(s^2) = \sqrt{s^2 + \varepsilon^2}$, where ε is a small fixed constant to avoid numerical problems. A linear approximation of the nonlinear Euler-Lagrange equations is created. This system of linear equations can be solved using SOR or a similar solver. To try to avoid all the local minimums that exist, this is embedded in some coarse-to-fine framework. While this sounds very similar to the previous linearization approach it is subtly different in that the linearization occurs while solving the original nonconvex problem Eq. 1.1. The previous approach instead solves a linearized version of the original problem.

In the first case we solve Eq. 1.2 exactly which we must remind ourselves is not the original problem. In the nonconvex case we approximate a solution to the original problem Eq. 1.1. Since it is nonconvex one does not know if they have found a global solution or indeed if there is a unique optimal solution. A multi-grid method is often used which moves from the coarse scale to the fine and back up to the coarse scale repeatedly in an effort to avoid falling into local minima.

1.4   Occlusions

Color constancy assumes that a point in the reference image appears in the target image. This assumption is violated each time one object moves in front of another or disappears from the frame altogether. The *occluded* pixels are ones that are hidden in the target image while *dis-occluded* pixels are uncovered in the target frame with no match in the reference frame.

Most optical flow methods ignore occlusion. This works to some extent because the occluded portion of the image is usually small. Furthermore, a robust color constancy penalty can absorb the occlusion errors while still getting much of the remainder correct.

Better results are generally possible if occluded pixels are correctly identified and masked from the color constancy error.

1.5   Total Variation

As a regularizer for optical flow, total variation sums up the absolute value change in flow from one pixel to the next. Similar to the evolution of the color constancy penalty, Horn and Schunck's work [4] used a quadratic penalty on the magnitude of the flow gradient. Changing to an L^1 penalty of the gradient, or total variation as a regularizer, has had a dramatic improvement on optical flow quality. The total variation penalty allows edges while still promoting smoothness in the flow.

1.6 From Optical Flow to Trajectory

Trajectories represent the first topic in this section that is not present in 2-frame optical flow methods. Trajectories are formed by projecting the 3-D motion of an object across time onto a 2-D camera plane.

Trajectories differ from optical flow. Optical flow starts from a pixel grid of each image and points generally in between the grid in the next image. Trajectories may be anchored to the pixel grid for only one frame of the sequence. Optical flow is occluded if the pixel is not found in the next frame, a trajectory may be occluded for multiple frames and then resurface. Because trajectories have a one-to-one relationship to the physical world, under some circumstances the trajectories may exist in a low rank subspace similar to the 3-D objects themselves. We examine these differences in depth in Sect. 2.4.

Converting from optical flow to trajectories is non-trivial (see Sect. 2.2.3). Connecting broken tracks across frames is an under-constrained inverse problem requiring regularization to complete.

1.7 Sparsity

If one knows something of the structure of a signal it may be possible to represent it with just a few basis vectors. For example, a musical tone requires a dense set of samples to represent it in the time domain but typically requires only a few components of a Fourier basis. In the latter case the representation is said to be sparse since most of the elements are zero. In this example, if one knew the signal was a musical tone, then even in the presence of corruption, noise, or missing samples one still has an excellent chance of reconstructing the original signal using a sparsity constraint.

We demonstrate improvements in optical flow in Chap. 3 using sparsity as a regularizer.

1.8 Dictionary

Dictionary is used instead of basis when the number of vectors exceeds the dimension of the subspace. By definition, a basis has "just enough" vectors to span a subspace. It is usual that some combination of basis vectors occur repeatedly in a signal. By adding that combination to the dictionary, it now requires only one non-zero element in the representation instead of the previous combination. By adding that dictionary element, we have enabled a more sparse representation of the signal. Using this new dictionary allows a sparse constraint an even better chance of recovering an ambiguous signal.

Our hypothetical example illustrates that having a good dictionary is the key to getting a sparse representation. Good dictionaries can be learned from examples of representative signals like the signal one wants to represent. The best library is learned from the exact signal one wants to represent.

In other applications of computer vision like denoising, the dictionary can be trained on the noisy data. However, in the case of optical flow we do not have an *a priori* version of the flow. We illustrate in Chap. 3 a new method to learn the dictionary while the optical flow is computed.

1.9 Low Rank

It possible to write any matrix as a sum of rank one matrices. This can be seen by inspection of the *Singular Value Decomposition* (SVD) of a matrix. Suppose for simplicity A is $n \times n$ (A does not need to be square), then the SVD is

$$A = U \Sigma V^T = (u_1 \, u_2 \, \cdots \, u_n) \begin{pmatrix} \sigma_1 & & & \\ & \sigma_2 & & \\ & & \ddots & \\ & & & \sigma_n \end{pmatrix} \begin{pmatrix} v_1^T \\ v_2^T \\ \vdots \\ v_n^T \end{pmatrix} \tag{1.3}$$

$$A = \sigma_1 u_1 v_1^T + \sigma_2 u_2 v_2^T + \cdots + \sigma_n u_n v_n^T \tag{1.4}$$

where u_i and v_i are orthonormal vectors of length n. A is said to be of low rank if most of $\sigma_i = 0$ for $i = 1, 2, \ldots, n$. Low rank in a matrix is analogous to sparsity in a vector.

Irani [5] showed that trajectories of affine motion or orthographic camera models exist in a low rank subspace. This property has been used in these special cases but can not be applied to general motion with a perspective camera. In Chap. 4 we develop a new method using a robust coupling to a low rank constraint that improves trajectories in more general cases.

References

1. J. Aujol, G. Gilboa, T. Chan, S. Osher, Structure-texture image decomposition modeling, algorithms, and parameter selection. Int. J. Comput. Vis. **67**(1), 111–136 (2006)
2. T. Brox, J. Malik, Large displacement optical flow: descriptor matching in variational motion estimation. IEEE Trans. Pattern Anal. Mach. Intell. **33**(3), 500–513 (2011)

3. R. Garg, A. Roussos, L. Agapito, A variational approach to video registration with subspace constraints. Int. J. Comput. Vis. **104**, 286–314 (2013)
4. B. Horn, B. Schunck, Determining optical flow. Artif. Intell. **17**, 185–203 (1981)
5. M. Irani, Multi-frame correspondence estimation using subspace constraints. Int. J. Comput. Vis. **48**(153), 173–194 (2002)
6. S.E. Palmer, *Vision Science: Photons to Phenomenology* (The MIT Press, Cambridge, MA, 1999)
7. L. Rakêt, L. Roholm, M. Nielsen, and F. Lauze. TV-L 1 optical flow for vector valued images, in *Energy Minimization Methods in Computer Vision and Pattern Recognition*, pp. 329–343 (2011)

Chapter 2
Optical Flow and Trajectory Methods in Context

Abstract In this chapter we study the related fields of multi-frame optical flow and trajectories. Since the beginning of modern optical flow estimation methods, multiple frames have been used in an effort to improve the motion computation. We look at why most of these efforts have failed. More recently, researchers have stitched together sequences of optical flow fields to create trajectories. These trajectories are temporally coherent, a necessary property for virtually every real-world application of optical flow. New methods compute these trajectories directly using variational methods and low-rank constraints. We also identify the need for appropriate data sets and evaluation methods for this nascent field.

2.1 Introduction

Computing motion across a sequence of images is a fundamental computer vision task. Whether an autonomous vehicle seeks to avoid a collision or one is creating a special effect for cinema, identifying the tracks taken across time is crucial. To be useful, a track must be *temporally coherent*, an idea that is not yet achieved by most algorithms currently in use today.

Most optical flow methods are applied to two frames of a sequence in an attempt to map the motion from one frame to the next. These methods are then applied sequentially to pairs of images in a longer sequence. However, ambiguity generally exists because of occlusion, noise, lighting variations, and the general ill-posed nature of the problem. This confusion is easily propagated from one frame to the next creating a cascade of errors. One would rightly expect that looking across multiple frames would provide additional robustness.

Using multiple frames to improve optical flow is not a new idea. In fact, from the beginning of modern optical flow estimation methods, efforts were made to temporally smooth the flow. These early methods were not effective in practice, relying on an accurate locally computed temporal derivative which does not exist in the presence of typical motion. Recently published algorithms show promise but do not yet completely solve the problem, pointing to a huge research opportunity that has yet to be fully exploited.

© The Author(s) 2016

J. Gibson and O. Marques, *Optical Flow and Trajectory Estimation Methods*,
SpringerBriefs in Computer Science, DOI 10.1007/978-3-319-44941-8_2

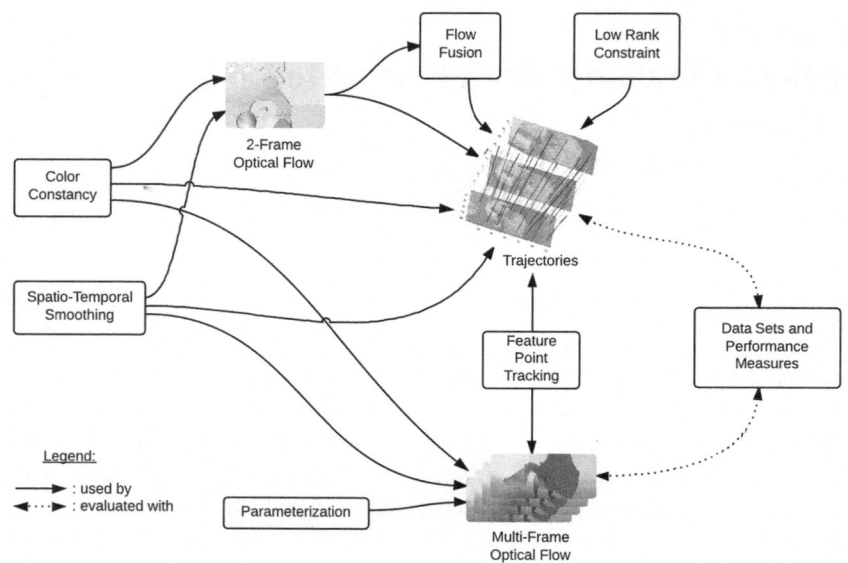

Fig. 2.1 Multi-frame optical flow and trajectory methodologies and relationships

This chapter examines two closely related areas, *multi-frame* methods and *trajectories*. Multi-frame methods use a sequence of images to compute one or more optical flow fields. This is a direct extension of optical flow in the temporal dimension. A trajectory, on the other hand, describes the motion of a particle across multiple frames. A dense set of trajectories describes the motion of the scene across a sequence of frames. We do not cover 2-frame optical flow methods. For excellent reviews of this field see Baker et al. [4] and Sun et al. [34].

We postulate that trajectories are the more desirable generalization of optical flow because of their temporal coherence. We also identify opportunities afforded by the lack of recognized trajectory data sets and evaluation methods.

Figure 2.1 shows a graphical abstract of this chapter. It is structured as follows: Sect. 2.2 is the heart of the chapter. It summarizes 30+ references in the field from an algorithmic point of view, reviewing algorithms from the last 30 years and placing them into categories.

Optical flow methods are often evaluated against publicly available benchmarks. Section 2.3 describes the most commonly used benchmarks and performance measures. Section 2.4 discusses the differences between optical flow and trajectories. Finally, Sect. 2.5 presents concluding remarks.

2.2 Algorithms

The goal of this section is to provide a representative list of optical flow methods organized by category and put into context. The selected algorithms and methods span 30+ years of research in this field. The categories here are more inspired by the chronological development of ideas rather than any mutually exclusive partition. There is an unavoidable overlap between the categories. For example, spatio-temporal smoothing is present in parameterization methods. Low-rank methods can be seen as a parameterization but are placed in their own section; one fusion flow method appears in the low rank category. When multiple methods are combined the most prominent or novel feature is used.

2.2.1 Spatio-Temporal Smoothing

There were many efforts to expand 2-frame optical flow estimation methods into multiple frames in the two decades following its "birth" with the parallel works of Horn and Schunck [18] and Lucas and Kanade [21] both in 1981. Since spatial gradients were being used to guide the flow in the 2-frame case there was a natural expansion into 3-D gradients of the spatio-temporal volume consisting of a sequence of frames stacked together.

Murray and Buxton [22] were concerned with solving segmentation using motion cues. They assume that the 3D scene they are considering is comprised of locally planar facets which greatly simplifies their motion model. They use an MRF to enforce a piecewise smooth spatial-temporal regularization and solve with simulated annealing. They assume that motion is temporally constant. This assumption, while a reasonable starting point is not supported in natural image sequences.

Heeger [17] proposes using 3D Gabor filters operating on a spatio-temporal volume. He operates on an image pyramid to find the scale which best matches the motion amount to his filters. Because the filters are relatively large, the results are poor with smaller objects or fine structure. Furthermore, the results are not valid near motion boundaries.

Nagel [23] proposes an extension of his image-driven smoothing across discontinuities into the spatio-temporal realm. He does not implement or show results but presents a theoretical framework to do so.

Singh [33], Chin et al. [11], and Elad and Feuer [13] all used a Kalman filter to enforce smoothness in the temporal domain.

Weickert and Schnörr [38] derive a flow-driven spatio-temporal smoothness constraint. Using a linearized color constancy in their objective, they solve the Euler-Lagrange equations.

Bruhn et al. [9] proposed a spatio-temporal version of their *Combined Local Global* algorithm. This algorithm combines the global nature of Horn and Schunck [18] with the local area matching of Lucas and Kanade [21].

Salgado and Sánchez [30] separate the spatial and temporal smoothing. Spatially they follows Nagel's [23] image-driven smoothing. Temporally they penalize any change in flow across frames. They solve Euler-Lagrange equations across a pyramid of images. Using two synthetic sequences with simple motion, they demonstrate improvement with their temporal method over spatial smoothing alone. According to [36], this temporal smoothing fails by oversmoothing complex motion.

Zimmer et al. [39] use a linearized version of the color constancy and then solve the Euler-Lagrange equations. For data constancy they use color and gradient of each HSV color channel each with a robust ℓ_1 approximate penalty. They point out limitations of image-driven regularization results in oversegmentation because every image edge does not correspond to a flow edge. They note flow-driven regularization does not give as clean flow edges as image-driven results. They combine the two using $||G\nabla u||_1$ where G is tensor-directed based on image gradient directions (image-driven) and the ℓ_1 norm is the flow-driven regularization. They observe that Middlebury sequences have too large of a displacement hence the assumption of a temporally smooth field is violated. So they use the *Marble* sequence with its slow motion to show spatio-temporal improvement over the 2-frame case.

Summary We observe that using a temporal smoothness as a constraint shows improvement over 2-frame methods for sequences with small frame-to-frame displacements. Unfortunately, for larger displacement, the locally generated temporal derivatives do not agree with the actual motion. This failure mode is in fact ordinary rather than the exception. Temporal sampling is far lower than the Shannon sampling rate and typical motion in fact is highly aliased. While spatial resolution is inherently bandlimited by camera optics, there is no natural mechanism which limits physical motion relative to the camera frame rate. This stumbling block has stifled development in this direction.

2.2.2 *Parameterizations*

In this subsection we present methods that make specific assumptions about the motion or camera model, then use this to predefine some basis functions.

Black and Anandan [6] assume that the image plane acceleration of a patch is approximately constant temporally and they construct a running average to constrain the variation with time. They further use a robust function to spatially smooth the flow while rejecting outliers, an improvement over the quadratic penalty which was commonly used at the time.

Black [5] improves his previous work to include a robust penalty on the constant acceleration assumption. Furthermore, he solves the problem with a continuation method called *Graduated Non-convexity* as a more deterministic strategy than the simulated annealing.

Volz et al. [36] uses a 5 frame sequence with a variational formulation. Their data constancy term separately penalizes the 4 incremental flows. Their objective functions sums the color constancy error across all 3 color channels. They also use a tensor-directed spatial regularizer. They experiment with applying the regularizer on each of the 4 flow fields separately and summing versus applying the tensor-directed regularizer to each flow field summing the result then applying an ℓ_1 approximation. The joint regularization seemed to generally work better than penalizing each separately.

They try to fit trajectories to a parabolic curve. This is based on their assumption that the projected 2D trajectories first derivatives are continuous. They compare just spatial regularization with 1st and 2nd order trajectory regularization to a locally adaptive version and a globally adaptive version. There was no clear winner. The parabolic trajectory assumption is similar to the constant acceleration assumption of [6].

They solve the Euler-Lagrange equations using a Gauss-Seidel multigrid solver in a coarse to fine framework. They make the point that they do not use a coarse-to-fine linearization of the problem or image warping. Because of this they outperform methods using a linear color constancy on large displacement motion. They instead linearize in the solution space via their multi-grid method. This involves "W" cycles of moving from coarse to fine several times to try to find a global minimum to a non-convex problem.

Nir et al. [24] use an over-parameterized space-time model to describe flow. Total variation regularization is applied to the coefficients of the basis functions. The basis functions are chosen to fit either an affine model, rigid-body motion, and special cases such as translational and constant motion. Euler-Lagrange equations are solved in a multi-scale framework. They showed improvements on very simple motion examples, *Yosemite*, *Flower Garden*, and a synthetic translational sequence. The basis functions consists of up to quadratic coordinate terms of pixel location chosen to fit the particular model. The spatio-temporal results were only reported on Yosemite.

Summary Simplifying the motion model allows a parameterization of the flow or trajectory. If the basis chosen can accurately represent the motion in the scene then the ill-posed flow problem is constrained, thereby improving results. Unfortunately, for general motion these models do not hold, limiting their usefulness to special cases.

2.2.3 Optical Flow Fusion

In this section we examine methods that begin with a sequence of 2-frame optical flow fields then fuse matching tracks into longer trajectories.

Sand and Teller [31] use a combination of sparse point tracking which extends over multiple frames and dense optical flow which is computed across two frames which they call *particle video*. The big difference between particle video and a dense trajectory is that the particle density is adaptive based on the motion detail so that fewer points have to be tracked.

Sand and Teller start with a KLT-based global motion stabilization step. They then compute 2-frame optical flow estimates for all the frames in the sequence. They identify occluded pixels and then mask the pixels from the color constancy error. They dryly point out that robust regularizers produce better looking transitions that are still wrong. They use Delaunay triangulation to link adjacent particles. Weights of edges are computed based on similarity of motion. Particles moving similarly will have strong links while links will be nearly zero across occlusion boundaries. Particles that have high distortion errors, a measure of how dissimilar their trajectory is to their neighbors, get pruned. Pixels whose color is scale invariant can get added as a particle to fill out areas of lower density. The number of particles tracked are much smaller than the number of pixels. This shorthand description of the image motion is computationally advantageous for matting, compositing and similar tasks. Sand and Teller's distortion measure and pruning can be seen as similar to the low rank DCT basis used later by Garg et al. [15], i.e. the assumption of high correlation between trajectories.

Rubenstein et al. [29] point out flaws in Particle Video (PV) [31] where frames far apart match a particle of a totally different color. Their work aims at achieving long motion trajectories and is not spatially dense. Interestingly, they use KLT as an initialization because they found it more stable than *Particle Video* or Brox and Malik's *Large Displacement Optical Flow* (LDOF) [8]. KLT continually drops tracks and restarts new ones with time. These authors use an MRF formulation to link temporally separated tracks. Along with a track compatibility measure they impose a track regularization which avoids crossing trajectories behind occlusions. Claiming no real benchmark for long term tracking existed they created some synthetic sequences. They show their methodology would improve PV and LDOF when they were used as initializers. They demonstrate the usefulness of these sparse long trajectories in human-action recognition problems.

Crivelli et al. [12] compute long-range motion by an iterative approach that works backwards from the reference frame N which is the last frame of the considered sequence. They look for the best path forward from the current frame n to frame N. They may choose from all the paths in frames between $n, \ldots, N$, to pick up a trail that may have been lost. In this way they are not forced to find a match in frame $n + 1$ when the match there may be occluded. They expand their backtracking approach to the symmetrical idea of considering frames after the reference frame which in turn is iteratively computed forwards from backwards-pointing flow. They use a Potts-style regularizer and solve for their optimal solution via graph cuts. They demonstrate some visual improvement over Sundaram and Brox [35].

Summary These flow fusion methods were the first to construct long trajectories while considering occlusion. Their emphasis is on finding the best way to connect disparate flow fields. Their results are limited by the quality of the flow field inputs.

2.2.4 Sparse Tracking to Dense Flow

Long term tracking of sparse points even with large displacement is a well developed topic, epitomized by the famous KLT tracker [32]. These trackers can often track objects across many frames although feature points get dropped when they can no longer be reliably matched and new ones are added. The algorithms in this subsection use feature point matching to try to address the large motion weakness of coarse-to-fine flow methods.

Brox and Malik [8] capture large motion using SIFT or HOG descriptors. This is then incorporated into a nonconvex variational model. They formulate the flow objective with a nonlinear optical flow constraint, using color and gradient constancy. The feature point matching flow is also used as a constraint but its influence is reduced to zero as the image is taken coarse to fine. Euler-Lagrange equations are then solved. Middlebury *Average Angular Error* is reported but the rest is qualitative comparison with other methods.

Sundaram et al. [35] is a GPU implementation of [8]. There are some GPU-specific adaptations from [8] including HOG descriptors and Conjugate Gradient solver with a pre-conditioner. They evaluate the second eigenvalue of the image structure tensor and discard any tracking points with the eigenvalue smaller than a threshold. Tracks are stopped when forward and backwards flow do not match. Additionally, tracks that are get very close to flow boundaries are stopped to prevent inadvertent drift to the other side of the boundary. The authors show a 66 % improvement against Particle Video [31] using Sand and Teller's data set. It also outperforms a KLT tracker in terms of accuracy and density. Because of the GPU implementation it also runs 78× faster than the CPU version.

Lang et al. [20] construct a method quite different from the other variational methods presented here. They pose the optical flow problem as a heat equation with discrete SIFT point matching becoming Dirichlet boundary conditions. By posing spatial smoothing as a quadratic penalty of the flow gradient à la Horn and Schunck, they recall that the solution is equivalent to Gaussian convolution. They apply this spatial filter within an edge-aware domain transform. Temporally they use a simple box filter along tracks. They trace a path or track forward from a reference frame until the track leaves the image or multiple tracks converge on one pixel. For pixels that do not have a track, a new one is started. In the case of multiple tracks on a single pixel, one is discarded at its previous frame. These techniques produce a temporally consistent result. Spatially, however, their smoothing method seems vulnerable to objects that lack sharp gradient boundaries to constrain the smoothing. Their positive results are mostly qualitative visual comparisons. The separable nature of the Gaussian filter enables their algorithm to execute quickly, about 64× faster than Volz et al. [36].

Summary By choosing tracking points that are distinctive across multiple frames, this strategy combines some of the best of two worlds. The high-confidence discrete point tracks are used to constrain variational optical flow to produce a dense flow field.

2.2.5 Low Rank Constraints

Under certain camera models or types of motion, the trajectories of a scene form a low rank matrix. We look at the research that leverages this property. Low rank trajectories constraint is not an extension of 2-frame optical flow but is only meaningful in the context of trajectories.

To illustrate a typical trajectory representation used by work in this section: Let N be the number of pixels in a frame and F be the number of frames in a sequence. Let u, v be the previously defined horizontal and vertical optical flow between a reference frame and every other frame in $1 \ldots F$. Here the flow fields have been vectorized and written as rows of the matrix

$$\mathcal{U} \triangleq \begin{bmatrix} u_{11} & \cdots & u_{1N} \\ \vdots & & \vdots \\ u_{F1} & \cdots & u_{FN} \\ v_{11} & \cdots & v_{1N} \\ \vdots & & \vdots \\ v_{F1} & \cdots & v_{FN} \end{bmatrix}. \tag{2.1}$$

The ith column of this matrix represents the trajectory of the ith pixel in the reference frame.

Brand [7] infers a 3D non-rigid model and its motion from a single video camera with a weak perspective camera assumption.

Irani [19] considers camera and motion models that result in a low rank trajectory matrix. Her work is derived from work in Structure from Motion (SfM) but she does not do any 3D reconstruction in this work. Further, she computes a dense set of trajectories for each pixel in a reference frame as in Eq. 2.1. This is different from most SfM work which requires a set of matched points that appears in each frame. She uses the posterior inverse covariance matrix of the flow as a confidence measure. These confidence measures ensure the low rank flow computed is propagated to regions of low confidence rather than allowing the low confidence regions to corrupt high confidence areas. Specifically, Irani shows how this low rank constraint applies to linearized camera models, such as orthograghic, weak-perspective, and to full perspective cameras undergoing small rotation and forward translation.

Garg et al. [14] use "reliable" tracks to compute a trajectory basis. They use an L^2 norm for color constancy and spatial smoothing. They use sequences about 50 frames long with only 5 basis elements. That is to say the column space of Eq. 2.1 is spanned by these 5 vectors. They parameterize the trajectories with these basis functions. In a later paper [15] they called that a hard constraint compared to their later soft constraint. They form the Euler-Lagrange equations and use a coarse-to-fine framework with linearization and warping to create linear equations solved with the SOR solver.

Ricco and Tomasi's *Lagrangian Motion Estimation* (LME) [25] combine occlusion and low rank multi-frame trajectories. They first use a standard tracker on key points and derive a trajectory basis from these. Trajectories in general are not of low rank but they claim the rank is often small in practice. They use the first and last frame of a sequence as reference frames. They reason that using the first frame is fine for points that are not occluded there but useless otherwise. Clearly, using only first and last does not solve this problem but mitigates it somewhat. They use a nonlinear color constancy and an image-driven TV regularizer. They have an occlusion mask variable v which they call a visibility flag. They apply two energy functions to v, the first is an L^2 penalty for v differences from v^0, a starting per-trajectory initialization. Secondly, they enforce a TV smoothness constraint which is relaxed when the trajectories change direction or the second temporal derivative of the brightness constancy error is high. The visibility flag allow entries in $\mathcal{U}$ which are low-rank constrained but do not contribute to the color constancy error. They only declare a trajectory occluded if they can find the occluding track. They visually outperform LDOF for trajectories that track heavily occluded areas, e.g. a lady walking in front of scene.

Garg et al. [15] implement a low-rank trajectory soft constraint with an anisotropic regularization. This is implemented in CUDA on an NVIDIA GPU. They created a trajectory benchmark based on a synthetic flag waving sequence. They outperformed LDOF and ITV-L^1 by Wedel et al. [37]. Their objective function contains a Huber-L^1 robust color constancy term with a Huber-L^1 approximation of Total Variation of the basis coefficients. They have an L^2 soft coupling term between their color constancy trajectory and the low-rank generated flow $\mathcal{U}$. They show the soft coupling produces better results than their previous "hard coupling". This can somewhat be explained since they do not use any occlusion masking. Results are shown for PCA basis trajectories computed from KLT tracks as well as a pre-defined DCT basis. The PCA performs better all although not always by a lot. One may consider these trajectories as a best fit low rank regression to the frames where the points are visible. In one way this is a reasonable approximation of the physical 3D path. The physical trajectory does not vanish just because it is not visible in a frame or two.

Ricco and Tomasi [27] significantly refine their previous model tying together low-rank trajectories and occlusions. Their objective consists of a data constancy and a temporal smoothing term. The data term is an L^1 approximation of a nonlinear color constancy. Data constancy is masked in occluded areas. Temporal smoothing is achieved by constraining trajectory coefficients that describe nearby paths through similar appearance to have similar values. Their visibility flag v is formulated as a Markov Random Field and solved concurrently with the trajectories using graph cuts.

Initially, KLT sparse tracking is performed. They apply a two-frame optical flow method to sequential pairs to assemble tracks similar to Sundaram et al. [35]. These tracks are used to find a basis. This process is described by Ricco and Tomasi [26].

Anchor points start in the first frame, for tracks that begin there. For tracks that end in last frame, anchor points are placed there. For tracks that begin or end at frames in between, anchor points may be placed in those frames, particularly for areas that are occluded in the first or last frame. Anchor points are not placed in a thin barrier around motion boundaries.

Optimization alternates between solving a continuous optimization for the trajectory coefficients given a fixed set of visibility flags. Then with trajectory fixed the combinatorial visibility flags problem is solved via graph cuts. Anchor points are added until each of their pixels have a track within one pixel and invisible trajectories are removed. Optimization stops when trajectories change less than a pixel in every frame.

From their discussion, it is implied that they perform poorly on highly deformable objects like Garg's flag sequence and on crowds. Their model is susceptible to lighting changes over these long sequences. They log a computation time of 150 hours for 60 frames on the *marple1* sequence.

They outperform their previous LME results as well as LDOF. In the absence of a benchmark, they measure how closely a trajectory maintains color constancy. They also measure mean path length and pixel distance to nearest track. They dramatically beat LDOF on the pixel distance and soundly improve on their previous LME method. Oddly, they lose on path length to LME on every sequence!

Ricco points out in [28] that MPI-Sintel's data set comes closest to being useful with their longer sequences and ground truth optical flow for each frame but argues this is still not useful for long trajectories since they do not follow any long term correspondences.

Summary Low rank constraint is undoubtedly a powerful tool for orthographic camera models or weak perspective models. One researcher applied the low rank constraint to more general cases and found that it can be helpful. However, a full perspective camera with large z-motion can produce trajectories that are far from low rank. So the usefulness of this technique for general motion is unclear.

2.3 Data Sets and Performance Measurement

2.3.1 Existing Data Sets

The goals of this section are to remind the reader (1) This field is highly driven by comparing new methods against publicly available popular benchmarks. (2) Middlebury [3, 4] is by far the most popular of the benchmarks. (3) Middlebury is a combination of sequences, ground truths, and performance measures. Despite its popularity it is not suitable for multi-frame for these reasons:

1. The Middlebury evaluation set has up to eight frames in a sequence but provides a ground truth for only the center pair.
2. For the low rank methods, eight frames form too short of a sequence to add a meaningful constraint.
3. No trajectory ground truth.

To fill this gap different data sets and performance measures have been proposed. This section discusses some of them. Middlebury, while venerable, is now considered largely solved and lacking challenges of realistic sequences. The most prominent new contenders are the KITTI benchmark and evaluation [2, 16] and the MPI-Sintel dataset [1, 10]. Interestingly, top performers on Middlebury have been seen to not perform well on KITTI and MPI-Sintel. KITTI offers a multiview version of its data set, with 20 frames per sequence. Ground truth however, is provided for the center frame only. The MPI-Sintel data set is the most promising, providing sequences 20–50 frames long with 2-frame optical flow ground truth between each pair of frames.

There are almost no publicly available data sets for trajectory evaluation. This has led to each research effort choosing different sequences and metrics for evaluation. This makes it difficult to compare the performance of different approaches.

For 2-frame optical flow measurement methods see [4]. These metrics have been applied with multi-frame methods which generate 2-frame flow results. In broad strokes, they provide three measures. The Average Endpoint Error and Average Angular error provide two ways of comparing the average difference of a field of 2-D vectors. According to Baker et al. [4], the Average Endpoint Error is the preferred metric of the two. The third metric is specific to image sequences. Suppose we capture a series of high speed photographs. If we compute optical flow correctly between a reference image and the second image following, the flow should be capable of accurately predicting the intermediate image. Since the intermediate image is known we compute an RMS error between our interpolated frame and the captured ground truth. For difficult sequences a visual qualitative inspection is more informative than numbers alone. An extended version of this methodology is often used with trajectories to warp sequences back to a reference frame.

2.3.2 Individual Measurement Efforts

Sand and Teller [31] took video sequences and appended the reverse order of the same frames. They admit this should not be used for future comparisons. The only known here is that the last frame is the same as the first, so the zero vector satisfies this requirement without following any meaningful trajectory in the intermediate frames.

For qualitative comparison Ricco and Tomasi [27] use the trajectories to warp all frames to a reference frame. Quantitatively, they measure color constancy along a trajectory masked by occlusion. To avoid gaming this admittedly weak metric, they also measure the average visible length of the tracks. While color constancy and long visible tracks are desirable traits, neither directly imply accuracy. They also measure the distance from every pixel in the sequence to the nearest visible track to demonstrate the density of their implementation. However their implementation adds tracks until each of their pixels have a track within a one pixel.

Garg et al. [15] warps real-life sequences and add textures to show qualitative improvement. For quantitative improvement they use a motion captured flag sequence to create an orthographic camera synthesized version with ground truth. They have made this available on their web site. The sequences used seem to be selected such that trajectories do not leave the raster.

As trajectory algorithms improve, sequences with a perspective camera model and with content entering and leaving the image will be necessary.

Summary The path to recognition for any new 2-frame optical flow method runs through one of the publicly available evaluation sites. An opportunity awaits for a similarly useful data set and evaluation site for trajectory algorithms.

2.4 Trajectory Versus Flow

In this section we argue that trajectories are fundamentally different, more useful, hence more valuable, than sequences of 2-frame flow fields.

Multi-frame methods offer the possibility of improving 2-frame flow methods by applying additional constraints. Trajectory representation brings a temporally coherent motion that is desirable in most applications and necessary in others.

The 2-frame flow methodology grew as a 2-D extension of the 1-D stereo depth problem. Indeed, the Middlebury optical flow set contains stereo pairs as 1-D flow sets. However, unlike stereo data sets where additional cameras will not be added *ex post facto*, the premise of video is the existence of additional frames. It is difficult to think of an uncontrived application where optical flow computation stops after two frames. Perhaps the strongest legacy reason for 2-frame methods is their smaller computational requirements, in particular smaller memory.

Because of the ill-posed nature of optical flow, any method is presumed to contain errors. Simply concatenating these errors across frames quickly results in wildly inaccurate trajectories. This leaves the application relying on the flow, to somehow smooth the temporal inconsistencies. This effort, clearly is better conducted in the environment where flow constraints are present and can be managed.

To illustrate the importance of temporal consistency we consider a few ordinary applications of optical flow.

Optical flow is used in media and entertainment applications to add special effects, compositing, de-interlacing, speed changes, denoising and restoration. Since the end result is video or a cinema sequence, there must be temporal coherence. It is easier to tolerate small, temporally consistent errors here than similarly small uncorrelated errors which appear as frame-to-frame jittery noise.

Optical flow can be used to help disambiguate other inverse computer vision problems such as stereo depth (scene flow), segmentation, denoising, and structure from motion.

In robotic applications optical flow provides feedback on ego-motion as well as environmental hazards. Indeed, the KITTI evaluation benchmarks are street scenes captured from a moving vehicle. The requirements here for real-time, low-latency feedback provide a constraint on the number of look-ahead frames that are useful for a multi-frame scheme.

A sequence of two frame optical flow fields does have an advantage over a trajectory when warping an image. That is, optical flow provides a map from the center of each pixel, ignoring occlusion. Trajectories, on the other hand, may be anchored on a reference pixel in one frame and miss pixel centers in all the others. Of course, the value of the flow at a pixel center can be interpolated, or conversely with dense trajectories the color at the trajectory center can be interpolated.

It would be useful to convert trajectories to 2-frame flows or some intermediate length. This would ease comparisons not addressed by trajectory benchmarks. Additionally, it is convenient in image warping to have a flow or trajectory that is referenced to that image pixel centers. While methods have been introduced to fuse short flows to longer trajectories, the opposite has not been studied.

Summary Applications that use optical flow are ultimately interested in estimating the temporal consistency of trajectories. Trajectory computation is typically more expensive than 2-frame flow fields. Faster computers and better optimization methods are enabling technologies to the implementation of emerging trajectory estimation methods.

2.5 Conclusions

Since near the beginning of modern optical flow estimation methods, there have been efforts aiming at improving flow results with knowledge from additional frames. The first two decades of research focused on methods using the local computation of temporal derivatives. We have discussed how this endeavor fails in general. Research in the last 15 years has moved beyond color constancy to use low rank constraints and optimized bases to describe trajectories. We argue that 2-frame optical flow is largely a stop-gap for trajectories with their inherent temporal consistency. Algorithm development for dense trajectories is still immature. Similarly, the field lacks a recognized data set and evaluation methodology. We expect to see development along these three avenues — algorithms, data sets, and evaluation methods — to be further advanced in the next few years.

References

1. MPI Sintel Flow Dataset. http://sintel.is.tue.mpg.de/ (2014)
2. The KITTI Vision Benchmark Suite. http://www.cvlibs.net/datasets/kitti/index.php (2014)
3. The Middlebury Computer Vision Pages. http://vision.middlebury.edu (2014)
4. S. Baker, D. Scharstein, J.P. Lewis, S. Roth, M.J. Black, R. Szeliski, A database and evaluation methodology for optical flow. Int. J. Comput. Vis. **92**(1), 1–31 (2011)
5. M. Black, Recursive non-linear estimation of discontinuous flow fields, in *Computer Vision—ECCV'94*, pp. 138–145 (1994)
6. M.J. Black, P. Anandan, Robust dynamic motion estimation over time, in *Proceedings of Computer Vision and Pattern Recognition, CVPR-91*, pp. 296–302 (1991)

7. M. Brand, Morphable 3D models from video, in *Computer Vision and Pattern Recognition*, pp. II–456–II–463, vol. 2 (2001)
8. T. Brox, J. Malik, Large Displacement optical flow: descriptor matching in variational motion estimation. IEEE Trans. Pattern Anal. Mach. Intell. **33**(3), 500–513 (2011)
9. A. Bruhn, J. Weickert, C. Schnörr, Lucas/Kanade meets Horn/Schunck: combining local and global optic flow methods. Int. J. Comput. Vis. **61**(3), 211–231 (2005)
10. D.J. Butler, J. Wulff, G.B. Stanley, M.J. Black, A naturalistic open source movie for optical flow evaluation, in *European Conference on Computer Vision (ECCV)*, pp. 611–625 (2012)
11. T.M. Chin, W.C. Karl, A.S. Willsky, Probabilistic and sequential computation of optical flow using temporal coherence. IEEE Trans. Image Process. (A Publication of the IEEE Signal Processing Society). **3**(6):773–788 (1994)
12. T. Crivelli, P. Conze, P. Robert, P. Pérez, From optical flow to dense long term correspondences, in *International Conference on Image Processing (ICIP)*, pp. 61–64 (2012)
13. M. Elad, A. Feuer, Recursive optical flow estimationadaptive filtering approach. J. Vis. Commun. Image Represent. **9**(2), 119–138 (1998)
14. R. Garg, L. Pizarro, D. Rueckert, L. Agapito, Dense multi-frame optic flow for non-rigid objects using subspace constraints. Comput. Vis. ACCV **2010**, 460–473 (2011)
15. R. Garg, A. Roussos, L. Agapito, A variational approach to video registration with subspace constraints. Int. J. Comput. Vis. **104**, 286–314 (2013)
16. A. Geiger, P. Lenz, and R. Urtasun, Are we ready for autonomous driving? the kitti vision benchmark suite, in *Conference on Computer Vision and Pattern Recognition (CVPR)* (2012)
17. D. Heeger, Optical flow using spatiotemporal filters. Int. J. Comput. Vis. 279–302 (1988)
18. B. Horn, B. Schunck, Determining optical flow. Artif. Intell. **17**, 185–203 (1981)
19. M. Irani, Multi-frame correspondence estimation using subspace constraints. Int. J. Comput. Vis. **48**(153), 173–194 (2002)
20. M. Lang, O. Wang, T. Aydin, Practical temporal consistency for image-based graphics applications. ACM Trans. Graph. **31**(34):31:4–31:8 (2012)
21. B.D. Lucas, T. Kanade, An iterative image registration technique with an application to stereo vision, in *Proceedings of Imaging Understanding Workshop*, pp. 121–130 (1981)
22. D. Murray, B.F. Buxton, Scene segmentation from visual motion using global optimization. Pattern Anal. Mach. Intell. IEEE Trans. PAMI. **9**, 220–228 (1987)
23. H.H. Nagel, Extending the 'oriented smoothness constraint' into the temporal domain and the estimation of derivatives of optical flow, in *Proceedings of the First European Conference on Computer Vision*. (Springer New York, Inc., 1990), pp. 139–148
24. T. Nir, A.M. Bruckstein, R. Kimmel, Over-parameterized variational optical flow. Int. J. Comput. Vis. **76**(2), 205–216 (2007)
25. S. Ricco, C. Tomasi, Dense lagrangian motion estimation with occlusions, in *2012 IEEE Conference on Computer Vision and Pattern Recognition*, pp. 1800–1807 (2012)
26. S. Ricco, C. Tomasi, Simultaneous compaction and factorization of sparse image motion matrices. Comput. Vis. ECCV (2012)
27. S. Ricco, C. Tomasi, Video motion for every visible point, in *International Conference on Computer Vision (ICCV)*, number i (2013)
28. S.M. Ricco, Video motion: finding complete motion paths for every visible point, in *ProQuest Dissertations and Theses*, p. 142 (2013)
29. M. Rubinstein, C. Liu, W. Freeman, Towards longer long-range motion trajectories, in *Proceedings of the British Machine Vision Conference* (2012)
30. A. Salgado, J. Sánchez, Temporal constraints in large optical flow estimation, in *Computer Aided Systems Theory EUROCAST 2007*, vol. 4739, Lecture Notes in Computer Science, ed. by R. Moreno-Díaz, F. Pichler, A. Quesada-Arencibia (Springer, Berlin, 2007), pp. 709–716
31. P. Sand, S. Teller, Particle video: long-range motion estimation using point trajectories. Int. J. Comput. Vis. **80**(1), 72–91 (2008)
32. J. Shi, C. Tomasi, Good features to track, in *IEEE Conference on Computer Vision and Pattern Recogniton 1994* (1994)

33. A. Singh, Incremental estimation of image-flow using a kalman filter, in *Proceedings of the IEEE Workshop on Visual Motion*, pp. 36–43 (1991)
34. D. Sun, S. Roth, M.J. Black, A quantitative analysis of current practices in optical flow estimation and the principles behind them. Int. J. Comput. Vis. **106**(2), 115–137 (2013)
35. N. Sundaram, T. Brox, K. Keutzer, Dense point trajectories by GPU-accelerated large displacement optical flow, in *Computer Vision ECCV 2010* (Springer Berlin Heidelberg, 2010)
36. S. Volz, A. Bruhn, L. Valgaerts, H. Zimmer, Modeling temporal coherence for optical flow, in *2011 International Conference on Computer Vision (ICCV)*, pp. 1116–1123 (2011)
37. A. Wedel, T. Pock, C. Zach, H. Bischof, D. Cremers, An improved algorithm for TV-L 1 optical flow, in *Statistical and Geometrical Approaches to Visual Motion Analysis*, pp. 23–45 (Springer Berlin, Heidelberg, 2009)
38. J. Weickert, C. Schnörr, Variational optic flow computation with a spatio-temporal smoothness constraint. J. Math. Imaging Vis. pp. 245–255 (2001)
39. H. Zimmer, A. Bruhn, J. Weickert, Optic flow in harmony. Int. J. Comput. Vis. **93**(3), 368–388 (2011)

Chapter 3
Sparse Regularization of TV-L^1 Optical Flow

Abstract Optical flow is an ill-posed underconstrained inverse problem. Many recent approaches use total variation (TV) to constrain the flow solution to satisfy color constancy. In this chapter we show that learning a 2-D overcomplete dictionary from the total variation result and then enforcing a sparse constraint on the flow improves the result. We describe a new approach that uses partially-overlapping patches to accelerate the calculation and is implemented in a coarse-to-fine strategy. Experimental results show that combining total variation and a sparse constraint from a learned dictionary is more effective than employing total variation alone.

3.1 Introduction

In this chapter we use total variation to estimate flow, from which overlapping patches are created. We learn an overcomplete dictionary which allows a sparse representation of these patches. By *sparse* we mean that almost all of the vector entries are zero. Intuitively speaking, it means we can represent any flow patch for a given image sequence accurately with only a few patches from the dictionary. An overcomplete dictionary is distinguished from a basis in that it has more elements than the dimensionality it seeks to span. This encourages the representation to be more sparse than if we had just enough elements to span the space. In our approach, the dictionary is learned from the actual sequence instead of learning ground-truth flows using the leave-one-out approach of [12].

Our experimental results (Sect. 3.4) show that learning an overcomplete dictionary and combining a sparseness penalty improves flow results over total variation alone. Moreover, the computation of flow patches with less overlap dramatically reduces computational complexity without sacrificing accuracy.

The intent is not to produce the highest ranking algorithm on the Middlebury [1] evaluation site but to take well-known TV algorithms and show that sparse coding improves their result over a suite of test sequences.

© The Author(s) 2016   25
J. Gibson and O. Marques, *Optical Flow and Trajectory Estimation Methods*,
SpringerBriefs in Computer Science, DOI 10.1007/978-3-319-44941-8_3

3.2 Previous Work

In this section we look at some highlights in the evolution of optical flow methods that have influenced our work. For a general overview of the field we refer readers to surveys like [4, 18].

Most methods on the Middlebury Flow evaluation site [1] have some roots in Horn and Schunck's seminal paper [11]. Horn and Schunck posed a global description of optical flow in which they minimize the ℓ_2 norm of the color constancy error balanced against flow variation. The variational constraint serves to smooth and propagate the flow over homogeneous areas.

These two fundamental elements of flow estimation remain the foundation of modern methods. Brox et al. [6] made a big improvement to this model by replacing the ℓ_2 norm with ℓ_1. For the flow this implements total variation, which preserves natural discontinuities while still enforcing smoothness. Furthermore, color constancy constraint is prone to large errors due to noise, lighting change, reflections or occlusions. So additional improvements are seen from using the more computationally challenging ℓ_1 norm on the color constancy term as well.

The use of gradient constancy can help improve the natural inconsistencies in color constancy due to lighting or specular highlights. Using ROF [2, 19] structure and texture decomposition on the input image achieves similar results without the overhead of the additional gradient constancy term in the objective.

Total variation constraint of the optical flow is remarkably effective for its simplicity. It is however a blunt instrument. Despite its ability to promote smoothness while permitting discontinuities it doesn't have the capacity to capture additional structure of the flow.

There has recently been some exploration into using sparse representation as a regularizer of optical flow. Some of the most representative efforts are summarized next.

Shen and Wu [16] used Haar wavelets as a basis for the flow, enforcing sparse coefficients on a single small-scale image. They worked with quarter-size images from Middlebury and showed improved results over simplistic flow models.

Jia et al. [12] learns overcomplete dictionaries from ground-truth flow fields [4]. They make the computationally-easing assumption that flow can be learned with separate horizontal and vertical dictionaries. From their learned dictionary a sparse representation of the flow solution is sought. Their work does show that with a great dictionary, in their case ground-truth-based, sparse solutions are effective regularizers of color constancy. In our work, however, we learn an overcomplete dictionary from the flow we are computing, that is, we do not need or use ground truth in our dictionary computation.

Ayvaci et al. [3] and Chen et al. [7] have done recent work concerning sparsity and optical flow. The term sparsity in their research refers to color constancy errors. They are not attempting a sparse representation of the flow. Other than by lexicon, their work is not related to ours.

3.3 Our Work

The work described in this chapter was inspired in part by the image denoising success of Elad and Ahron [9] and Mairal et al. [14]. They broke the image into maximum overlapping patches then trained an overcomplete dictionary from those patches. By representing the patches with only a sparse number of dictionary elements the structure of the image was reproduced without the noise. In this sense we want to capture flow structure in a dictionary and use a sparse representation of these flow patches to add regularization to traditional flow estimation. However, differences quickly emerge since the flow we want to learn is not given as the image was in the denoising template. This implies there must be some bootstrap step to construct a flow approximation which can be learned and then iteratively refined.

We'll now look in more depth at each of these steps.

3.3.1 Partially-Overlapping Patches

Looking for sparse representations of patches stems from the reality that image-sized signals are too large to look for sparse representations from a computational tractability standpoint. The assumption then is that by studying patches we will find that the structure of the signal will be some sparsely representable lower dimensional signal embedded in the higher dimension of the image space. The computational complexity of sparse learning and representation is a combinatorial function of the patch size, dictionary size, and number of patches. So we are motivated to find the smallest patches and smallest number of patches that effectively capture the structure of the flow.

Given an image with P patches of size n, a $\beta\times$ overcomplete dictionary $D \in \mathscr{R}^{2n \times 2\beta n}$ and a flow $v \in \mathscr{R}^{2N \times 1}$, we want to find a sparse representation $a_{ij} \in \mathscr{R}^{2\beta n \times 1}$ for each patch. We can express this as

$$\hat{a}_{ij} = \arg\min_{a_{ij}} \mu ||a_{ij}||_0 + ||Da_{ij} - R_{ij}v||_2^2 \tag{3.1}$$

where $R_{ij} \in \{0, 1\}^{2n \times 2N}$ extracts the ijth patch from the flow and $\mu \in \mathscr{R}$ is a penalty weight of sparsity versus data fidelity. The patch and image have n and N elements respectively. This problem is NP-hard but approximate solutions exist. We use a maskable Orthogonal Matching Pursuit (OMP) algorithm [13]. This is a greedy algorithm that iteratively selects the atom from the dictionary with the largest projection onto the data set. The best least squares match of the vectors chosen so far is computed. That match is then subtracted from the data set and the process is repeated on the residue. See Algorithm 1 for details.

Input: We are given the matrix $A \in \mathbb{R}^{m \times n}$, the vector b, and the error threshold ε_0.
Output: $\min_x ||x||_0$ subject to $Ax = b$

1 $k \leftarrow 0$;
2 $x^0 \leftarrow 0$;
3 $r^0 \leftarrow b - Ax^0 = b$;
4 $S^0 \leftarrow \text{Support}(x^0) = \emptyset$;
5 **repeat**
6 $\quad$ $k \leftarrow k + 1$;
7 $\quad$ **forall the** $j \notin S^k$ **do** ; $\qquad$ // Find $\min_{z_j} ||a_j z_j - r^{k-1}||_2^2$
8
9 $\quad$ $\left| \quad \varepsilon(j) \leftarrow ||r^{k-1}||_2^2 - \frac{(a_j^T r^{k-1})^2}{||a_j||_2^2} \right.$;
10 $\quad$ **end**
11 $\quad$ $j_0 \leftarrow \arg_j \min \varepsilon(j)$;
12 $\quad$ $S^k \leftarrow S^{k-1} \cup \{j_0\}$;
13 $\quad$ $x^k \leftarrow \min_x ||Ax - b||_2^2$ subject to $\text{Support}(x) = S^k$;
14 $\quad$ $r^k \leftarrow b - Ax^k$;
15 **until** $||r^k||_2 < \varepsilon_0$;
16 Return x^k;

Algorithm 1: Orthogonal Matching Pursuit

In the denoising work of [9, 14] image patches are offset by one pixel horizontally and vertically. Reconstructing the image consists of averaging these patches together, thereby eliminating patch-sized blockiness. This is straightforward to implement but creates a large albeit sparse representation. Here the number of patches P is approximately the same the number of pixels N. In our work we consider 10×10 patches with a $4\times$ overcomplete dictionary. In this case the sparse representation a would be approximately 800 times larger than the original image. This becomes intractable for all but tiny images. We have observed useful results with cardinalities of 2–15, still producing quite large sparse structures.

The motivation for using partially overlapping patches stems from asking whether we can reduce the number of patches and hence complexity while not sacrificing accuracy. In Table 3.1 we look at the actual number of patches for an example image as we vary patch offset. We can see that the number of patches is approximately proportional to $1/k^2$ where k is the offset. There is clearly a large computational advantage to using values of $k > 1$.

As an experiment to evaluate the error introduced from partial overlap we learn a set of overcomplete dictionaries with 10×10 patches from each of the eight ground truth flows from Middlebury [1]. Then for each test sequence we create patches with offset $k = 1, \ldots, 10$. For this comparison we use a slightly different form in place of Eq. 3.1 where the data fidelity term is minimized and the cardinality is constrained. This is because at any patch we can achieve more accuracy by allowing the cardinality to increase. So for a meaningful comparison the cardinality was maintained constant for a given image sequence. We reconstruct the flow from the patches and measure the

Table 3.1 Comparing number of patches generated from a 480× 520 sized image versus patch offset

Offset	# Patches	Savings
1	217925	1×
2	54720	4×
3	24511	9×
4	13920	16×
5	8932	24×
6	6208	35×
7	4565	48×
9	2795	78×
10	2301	95×

The savings in computational complexity increases with the square of the offset

Average Endpoint Error (Avg EE) [4]. The average of all of these Average Endpoint Errors is plotted in Fig. 3.1. Although the individual results, not shown, are somewhat erratic, on average the error is very close to linear with respect to offset.

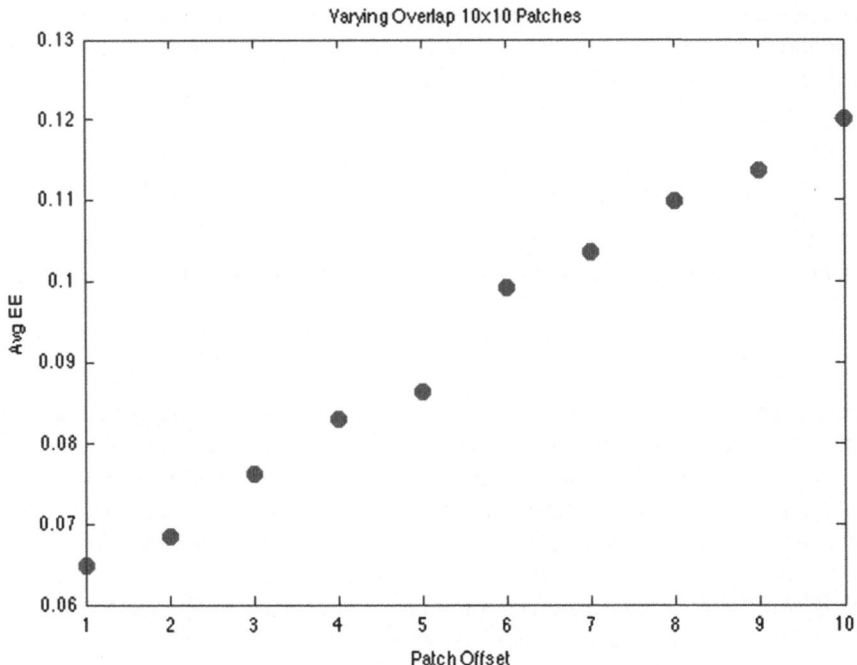

Fig. 3.1 Average EE for sparse patch reconstruction of Middlebury ground truths versus offset. Error is approximately linear with respect to offset

This suggests, as experiments later show, that partial-overlapping patches can reduce the computational complexity of sparse representation by an order of magnitude with minimal accuracy impact. In fact we show that partial overlap can actually improve results relative to full overlap.

3.3.2 Dictionary Learning

Consider the same variables used in Eq. 3.1 except now we want to create an ideal dictionary for sparse representation of the flow. This problem can be written as

$$\hat{D} = \arg\min_{D, a_{ij}} \sum_{ij} ||a_{ij}||_0$$

$$\text{subject to}\quad ||Da_{ij} - R_{ij}v||_2^2 < \varepsilon. \tag{3.2}$$

We find an approximate solution to this NP-hard problem using a Method of Optimal Directions (MOD) algorithm first described by Engan, et al. [10]. This iterative method discards the globally least used atom and adds an atom in the direction of the largest error. We modified an implementation by Elad [8] to allow data masking.

Figure 3.2 shows an example of a flow dictionary, whose elements are 2-D vectors color-coded in the Middlebury tradition.

3.3.3 Sparse Total Variation

A typical optical flow objective function using a total variation regularizer looks like

$$\arg\min_{v}\{||\rho(v)||_p + \lambda_{TV}(||Gv_1||_1 + ||Gv_2||_1)\} \tag{3.3}$$

where $\rho(v) = \nabla I \cdot v + I_t$, I is the image intensity, $v = (v_1, v_2)$ is the optical flow, I_t is the image temporal derivative, $G = (\partial_x, \partial_y)^T$, and λ_{TV} is the relative weighting of total variation penalty. The first term represents the color constancy penalty with an ℓ_p norm. As proposed by Brox et al. [6] and Wedel et al. [19], we consider $p = 1$ since the ℓ_1 norm more closely matches the long-tailed color constancy error.

We also consider another version from Werlberger et al. [20] where $G = D^{1/2}\nabla$ and $D^{1/2}$ is the diffusion tensor. This allows the flow crossing orthogonal to an image edge to change without penalty while smoothly increasing the penalty as the direction changes to parallel the image edge.

We add to Eq. 3.3 a sparsity penalty and a coupling term between the sparse representation and the flow.

Fig. 3.2 2-D learned flow dictionary for Rubber Whale, 8×8 patches, $2\times$ overcomplete

$$\arg \min_{v, a_{ij}} \left\{ ||\rho(v)||_1 + \lambda_{SP}(||Gv_1||_1 + ||Gv_2||_1) + \right.$$
$$\left. + \mu \sum_{ij} ||a_{ij}||_0 + \frac{\tau}{2} \sum_{ij} ||R_{ij}v - Da_{ij}||_2^2 \right\} \tag{3.4}$$

To make Eq. 3.4 easier to approximate we break it into two subproblems. First we assume that a is fixed which yields a convex problem in v. Secondly we fix v and solve for a. We iterate between these two subproblems a few times.

The first subproblem is described by

$$\arg \min_{v} \left\{ ||\rho(v)||_1 + \lambda_{SP}(||Gv_1||_1 + ||Gv_2||_1) + \right.$$
$$\left. + \frac{\tau}{2} \sum_{ij} ||R_{ij}v - Da_{ij}||_2^2 \right\}. \tag{3.5}$$

We solve this problem using Nesterov's method [15] described below. This algorithm allows an optimal convergence rate that scales with $\mathcal{O}(1/k^2)$ where k is the number of iterations even for this non-smooth objective.

The second subproblem is solved independently for each patch using OMP and has already been described in Eq. 3.1.

3.3.4 Nesterov's Method

Because the first-order Taylor approximation in $\rho(v)$ is only valid for small values of v, it requires the standard coarse-to-fine algorithm and multiple warps within a given pyramid level. At each call, we are solving for an incremental flow difference Δv between our existing flow estimation v_0 and our new estimation $v = v_0 + \Delta v$.

We first compute the gradient of Eq. 3.5. The smooth term is straightforward

$$\nabla_v \frac{\tau}{2} \sum_{ij} ||R_{ij}v - Da_{ij}||_2^2 =$$

$$\tau \sum_{i,j} R_{i,j}^T (R_{i,j}(v_0 + \Delta v) - Da_{i,j}) \tag{3.6}$$

$$= \tau (R^T R(v_0 + \Delta v) - \sum_{ij} R_{ij}^T a_{ij}). \tag{3.7}$$

The last form allows the calculation to be greatly accelerated by observing $R = \sum_{ij} R_{ij}$ and $R^T R$ do not depend on Δv and can be calculated in advance once per pyramid level. Furthermore, $\sum_{ij} R_{ij}^T a_{ij}$ can be calculated only once per warp and is in fact the reconstructed flow from the sparse representation. Therefore this otherwise slow patch-by-patch iterative calculation reduces essentially to a matrix multiplication and sum.

To compute the gradient for the non-smooth ℓ_1 norms we use the prox-function solution suggested by Becker et al. [5] which is

$$f_\alpha(x) = \max_{||u||_\infty \leq 1} \langle u, x \rangle - \frac{\alpha}{2} ||u||_2^2. \tag{3.8}$$

This is differentiable with the derivative

$$\nabla f_\alpha(x)_i = \begin{cases} \alpha^{-1} x_i & \text{if } x_i < \alpha \\ \text{sgn } x_i & \text{otherwise.} \end{cases} \tag{3.9}$$

The derivatives of the flow gradients and color constancy term follow easily. The Lipschitz constant is

$$L = \frac{8\lambda}{\alpha_{TV}} + \frac{||\nabla I^T \nabla I||_2}{\alpha_{Data}} + \tau ||R^T R||_2. \tag{3.10}$$

The last term is only present for the sparse-TV invocations of Nesterov's algorithm.

We initialize $v_0 = 0$, let $\Psi(v)$ be the objective function in Eq. 3.5 then the iterative equations for Nesterov's algorithm are:

Input: We are given input images $I(x, t_0)$, $I(x, t_1)$, patch size n, and offset k
Output: Optical flow v

1 Construct pyramid of images I^j for $j = 1, \cdots, pyramidLevels$;
2 $v \leftarrow 0$;
3 **for** $j \leftarrow pyramidLevels$ **down to** 1 **do**
4   **for** $i \leftarrow 1$ **to** $TVwarps$ **do**
5        $v \leftarrow \texttt{Nesterov}(\nabla I, I_t, v)$ Eq. 3.3;
6   Warp $I^j(x, t_1)$, calculate I_t;
7   **end**
8     $D^i \leftarrow OvercompleteDCT$;
9   **for** $i \leftarrow TVwarps + 1$ **to** $warps$ **do**
10       $flowPatches \leftarrow \texttt{FlowToColumn}(v, n, k)$;
11       $D^{i+1} \leftarrow \texttt{MOD}(flowPatches, D^i)$ Eq. 3.2;
12       $a \leftarrow \texttt{OMP}(v, D^{i+1})$;
13   **for** 1 **to** $iterations$ **do**
14          $v \leftarrow \texttt{Nesterov}(\nabla I, I_t, a, v, D^i)$ Eq. 3.5;
15          $a \leftarrow \texttt{OMP}(v, D^i)$ Eq. 3.1;
16   **end**
17   Warp $I^j(x, t_1)$, calculate I_t;
18   **end**
19 **end**
20 Return v;

Algorithm 2: Sparse Total Variation

$$\sigma_k = \frac{1}{2(k+1)}, \quad \gamma_k = \frac{2}{(k+3)} \tag{3.11}$$

$$y_k = v_k - L^{-1}\nabla\Psi(v_k) \tag{3.12}$$

$$z_k = v_0 - L^{-1}\sum_{i \le k}\sigma_i\nabla\Psi(v_k) \tag{3.13}$$

$$v_{k+1} = \gamma_k z_k + (1 - \gamma)y_k. \tag{3.14}$$

The individual pieces are shown assembled in Algorithm 2.

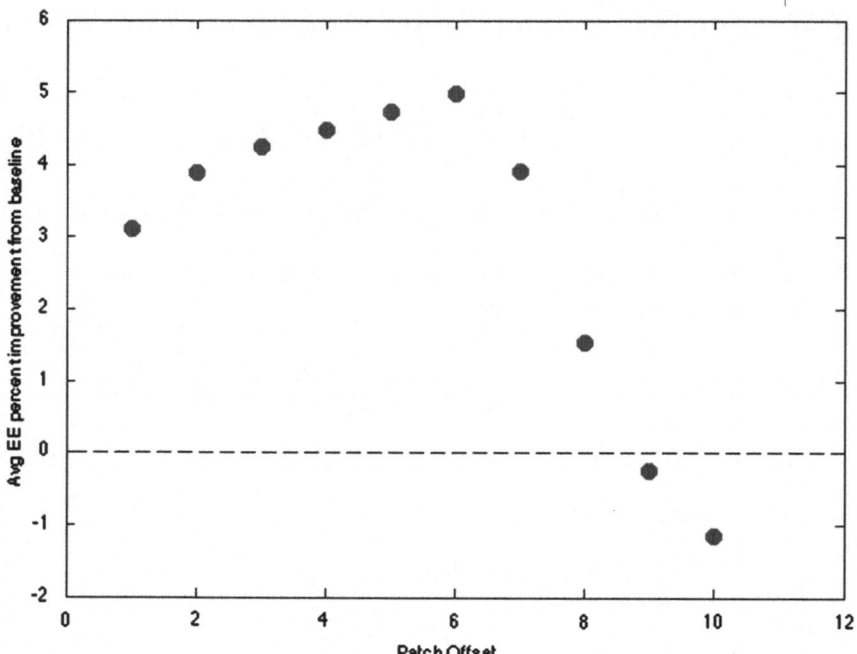

Fig. 3.3 Average percent improvement of Avg EE for 8 Middlebury sequences using tensor-directed 10×10 sparse total variation versus patch offset

3.4   Experimental Results

Experimental results are presented for four different dictionary types described below and two versions of the TV optical flow. Additionally, the average results of the tensor-directed version is shown for different patch offsets (Fig. 3.3).

3.4.1   Dictionaries

Experiments were conducted using the following dictionaries:

DCT
A predefined overcomplete DCT dictionary is used. This is the only instance where the dictionary is not learned and serves as a comparison for the effectiveness of a learned dictionary.

1D
Optical flow is a two dimensional vector. Patches of the horizontal and vertical elements of the flow are concatenated and a single dictionary is learned.

2×1D
Here two separate dictionaries are learned: one from the horizontal flow patches, the other from the vertical patches. This is the dictionary method used by Jia et al. [12].

2D
For 2D the horizontal and vertical patches are combined at each location, into a single patch of $2n$ elements. A single dictionary is then learned in this higher dimension.

The complexity and speed of computation is related to the number of dictionary elements. For that reason it seems meaningful to compare dictionaries of similar size. So the 4× overcomplete 2D dictionary is compared to the 8× overcomplete 1D dictionary because they both have the same number of dictionary elements.

3.4.2 Implementations Details

In the experiments shown for Middlebury images, the $pyramidLevels = 5$ and $TVwarps = 5$. For the coarsest two pyramid levels $warps = 5$, and for the finer levels $warps = 10$. The innermost loop constant in Algorithm 2 $iterations = 4$.

Table 3.2 shows the constants used in Eqs. 3.1–3.4. The OMP weighting constant $\mu = 0.05$ was used for all cases. Ten iterations of the MOD dictionary learning were run on each invocation with $\varepsilon = 0.001$. For the ℓ_1 proximal function in Eqs. 3.8 and 3.9, $\alpha_{TV} = \alpha_{Data} = 0.01$.

A standard ROF decomposition of structure and texture was performed on the input of all experiments and a 5×5 median filter was applied after each flow calculation as recommended by Sun et al. [17].

Table 3.2 Weighting constants used in our experiments

Pyramid Level	Simple TV		
	$100\lambda_{TV}$	$100\lambda_{SP}$	τ
1	5	2	0.16
2	1	0.5	0.02
3	0.5	0.5	0.02
4	0.3	n/a	n/a
5	0.06	n/a	n/a
	Tensor-directed		
1	16	4	0.24
2	1	0.5	0.03
3	0.5	0.5	0.03
4	0.3	n/a	n/a
5	0.06	n/a	n/a

In order to form a whole set of patches with offset k from an image it is necessary that $[h\ w] - [m\ n] = 0 \bmod k$, where h, w and m, n are the image and patch dimensions respectively. It may be necessary to augment the image by a small apron to maintain this relationship. The apron pixels are labeled with an out-of-bounds marker similar to the Middlebury ground truth occluded pixels. These values are masked off during the sparse learning and representation steps.

3.4.3 Discussion

The DCT dictionary performs worse than the baseline of no sparsity. It was seen, in results not shown, that sparse representations of the flow could be obtained with the DCT. While a sparse representation is a necessary condition for improvement, the DCT does not apparently capture the structure of the flow in a meaningful way. On the other hand we see that on the average, every learned dictionary method presented in Tables 3.3 and 3.4 outperforms the baseline.

Comparing Tables 3.3 and 3.4, as a percentage, sparse learning improves the simpler and slightly worse TV-only baseline more than the tensor-directed. We observe that while most of the learned dictionaries performed similarly for the TV-L^1 case, the 2-D dictionary performed noticeably better than the others in the tensor-directed case. In results not shown, we tried 8×8 size patches in the tensor-directed case and found improvement of about half as much as the 10×10 case but most noticeable was that the $2\times$ overcomplete 1-D dictionary performed best in that case. The trend would suggest that for larger patch size, the larger 2-D dictionary captures the structure better.

In Fig. 3.3 we compare the tensor-directed percentage Avg EE change relative to baseline averages for all possible offsets with 10×10 patches. Based on Fig. 3.1 increasing the patch offset should increase the error incurred, and it eventually does so, dramatically beyond $k = 6$. It is less obvious why smaller offsets make matters worse. Scrutiny of cases where it is worse has shown that averaging more patch approximations together creates both a smoother flow and more rounded edges, even where sharp angles are desired. So, for flow with large smooth areas, small patch offset will smooth the flow better than increasing the TV penalty. However, the errors incurred by rounding corners and edges often create a greater loss.

In addition to other structure, the sparse regularization 2-D dictionary model supports rigid-body translation. This may explain why sparse regularization at the coarser levels improves the flow results in the final level.

This algorithm may perform worse than baseline when at a coarser level the TV-only bootstrap flow makes a poor choice in an ambiguous area. This structure is later sometimes learned into the dictionary and then it is encouraged to persist by the sparseness penalty. In most but not all cases this is self-rectifying.

Table 3.3 TV-L[1] average endpoint error (Avg EE) for middlebury published ground truth set

Image Overcomplete dictionary size	TV-only	DCT 4×400	1D 4×400	2 × 1D 2×400	2D 2×400	DCT 8×800	1D 8×800	2 × 1D 4×800	2D 4×800
Dimetrodon	**0.1562**	0.1810	0.1782	0.1777	0.1766	0.1655	0.1775	0.1773	0.1774
Grove2	0.1921	0.2003	0.1743	**0.1727**	0.1775	0.2068	0.1760	0.1735	0.1769
Grove3	0.6869	0.7199	0.6489	0.6520	0.6504	0.7580	**0.6447**	0.6503	0.6519
Hydrangea	0.1597	0.1640	0.1572	0.1569	**0.1568**	0.1677	**0.1568**	0.1576	0.1580
RubberWhale	0.1118	0.1020	0.0993	**0.0984**	0.0985	0.1123	0.0988	0.0988	0.0993
Urban2	0.3801	0.5560	0.3440	0.3415	0.3430	0.4857	**0.3392**	0.3434	0.3406
Urban3	0.6441	0.6425	0.5382	0.5462	0.5405	0.8118	0.5513	0.5494	**0.5295**
Venus	0.2977	0.3587	0.2733	0.2736	**0.2715**	0.3556	0.2766	0.2717	0.2755
Avg % improved	0	−10.67	5.95	**6.08**	6.03	−12.84	5.82	5.94	5.92

Patch size is 10 × 10

Table 3.4 Tensor-directed TV-L^1 average endpoint error (Avg EE) for Middlebury published ground truth set

Image Overcomplete dictionary size	TV-only	DCT 4 ×400	1D 4×400	2 × 1D 2×400	2D 2×400	DCT 8 ×800	1D 8 ×800	2 × 1D 4 ×800	2D 4×800
Dimetrodon	0.1576	0.1630	0.1528	0.1514	0.1518	0.1626	0.1524	0.1516	**0.1512**
Grove2	0.2000	0.2060	**0.1757**	**0.1757**	0.1771	0.2025	0.1766	0.1781	0.1783
Grove3	**0.6630**	0.7447	0.6876	0.6929	0.6865	0.7415	0.6808	0.6868	0.6896
Hydrangea	0.1633	0.1686	0.1600	0.1605	0.1601	0.1678	0.1596	**0.1595**	0.1620
RubberWhale	0.1169	0.1174	0.1119	0.1115	0.1109	0.1171	0.1119	0.1120	**0.1103**
Urban2	0.3761	0.4946	0.3469	0.3541	0.3504	0.4806	0.3540	0.3531	**0.3427**
Urban3	0.6051	0.6887	0.5676	0.5719	**0.5626**	0.6776	0.5711	0.5737	0.5654
Venus	0.3035	0.3672	0.2815	**0.2757**	0.2817	0.3599	0.2831	0.2815	0.2818
Avg % improved	0	−11.09	4.87	4.80	4.96	−9.69	4.63	4.53	**4.99**

Patch size is 10 × 10

3.5 Conclusions

We have shown that flow structure can be learned from actual sequences in a boot-strap manner and used to further refine flow computation. Overcomplete dictionary learning and sparse flow representation have been demonstrated with generic total variation algorithms. The proposed method could easily be added to any more sophisticated variational approach. We have introduced a method of partial-overlapping patches that offers dramatic acceleration in the computation of this sparse representation.

Future research work includes using a computed occlusion mask [3] within the dictionary learning and sparse representation. Additionally, there seems promise in using a sparse model of rigid-motion in the image to constrain flow computation. The partial overlapping patch methodology developed here should also be useful in other patch-based applications such as denoising.

References

1. The Middlebury Computer Vision Pages. http://vision.middlebury.edu (2014)
2. J. Aujol, G. Gilboa, T. Chan, S. Osher, Structure-texture image decomposition modeling, algorithms, and parameter selection. Int. J. Comput. Vis. **67**(1), 111–136 (2006)
3. A. Ayvaci, M. Raptis, S. Soatto, Sparse occlusion detection with optical flow. Int. J. Comput. Vis. **97**(3), 322–338 (2011)
4. S. Baker, D. Scharstein, J.P. Lewis, S. Roth, M.J. Black, R. Szeliski, A database and evaluation methodology for optical flow. Int. J. Comput. Vis. **92**(1), 1–31 (2011)
5. S. Becker, J. Bobin, E. Candès, NESTA: a fast and accurate first-order method for sparse recovery. SIAM J. Imaging Sci. **91125**, 1–37 (2011)
6. T. Brox, A. Bruhn, N. Papenberg, J. Weickert, High accuracy optical flow estimation based on a theory for warping. Comput. Vis.-ECCV 2004, **4**(May), 25–36 (2004)
7. Z. Chen, J. Wang, Y. Wu, Decomposing and regularizing sparse/non-sparse components for motion field estimation, in *2012 IEEE Conference on Computer Vision and Pattern Recognition*, pp. 1776–1783 (2012)
8. M. Elad, *Sparse and Redundant Representations: From Theory to Applications in Signal and Image Processing*, 1st edn. (Springer Publishing Company, Incorporated, 2010)
9. M. Elad, M. Aharon, Image denoising via sparse and redundant representations over learned dictionaries. IEEE Trans. Image Process. **15**(12), 3736–3745 (2006)
10. K. Engan, S.O. Aase, J. Hakon Husoy. Method of optimal directions for frame design, in *Proceedings of the Acoustics, Speech, and Signal Processing*, pp. 2443–2446 (1999)
11. B. Horn, B. Schunck, Determining optical flow. Artif. Intell. **17**, 185–203 (1981)
12. K. Jia, X. Wang, X. Tang, Optical flow estimation using learned sparse model, in *Proceedings of IEEE International Conference on Computer Vision*, 60903115 (2011)
13. J. Mairal, F. Bach, J. Ponce, G. Sapiro, Online learning for matrix factorization and sparse coding. J. Mach. Learn. Res. **11**, 19–60 (2010)
14. J. Mairal, M. Elad, G. Sapiro, Sparse representation for color image restoration. IEEE Trans. Image Process. (A Publication of the IEEE Signal Processing Society) **17**(1), 53–69 (2008)
15. Y. Nesterov, Smooth minimization of non-smooth functions. Math. Program. **103**, 127–152 (2005)
16. X. Shen, Y. Wu, Sparsity model for robust optical flow estimation at motion discontinuities, in *2010 IEEE Conference on Computer Vision and Pattern Recognition (CVPR)*, vol. 1, pp. 2456–2463 (2010)

17. D. Sun, S. Roth, and M. Black. Secrets of optical flow estimation and their principles, in *2010 IEEE Conference on Computer Vision and Pattern Recognition (CVPR)*, pp. 2432–2439 (2010)
18. R. Szeliski. *Computer Vision: Algorithms and Applications*. (Springer, 2011)
19. A. Wedel, T. Pock, C. Zach, H. Bischof, D. Cremers, An improved algorithm for TV-L 1 optical flow, in *Statistical and Geometrical Approaches to Visual Motion Analysis*, pp. 23–45. (Springer Berlin, Heidelberg, 2009)
20. M. Werlberger, W. Trobin, T. Pock, A. Wedel, D. Cremers, H. Bischof, Anisotropic huber-L1 optical flow, in *Proceedings of the British Machine Vision Conference (BMVC)* (2009)

Chapter 4
Robust Low Rank Trajectories

Abstract In this chapter we show how sparse constraints can be used to improve trajectories. We apply sparsity as a low rank constraint to trajectories via a robust coupling. We compute trajectories from an image sequence. Sparsity in trajectories is measured by matrix rank. We introduce a low rank constraint of linear complexity using random subsampling of the data and demonstrate that, by using a robust coupling with the low rank constraint, our approach outperforms baseline methods on general image sequences.

4.1 Introduction

Since the beginning of modern optical flow methods, there has been the instinct that multiple frames should assist in the correct identification of ambiguous motion. Despite a substantial volume of work in the area, this goal has not generally been achieved.

In recent years, attention has shifted to *trajectories*, that is, a dense tracking of object points across multiple frames. The intrinsic temporal consistency of a trajectory makes it more attractive than a sequence of optical flow frames in virtually every application. Research on reliable methods for trajectory calculation is still in its early years, and several aspects associated with this field of research, such as recognized benchmarks, are yet to be fully developed. Additionally, most trajectory computational algorithms require much greater computation complexity than their optical flow counterpart.

To distinguish between optical flow and trajectories we note some differences. Optical flow can be loosely defined as the motion from the center of each pixel in a reference frame to its corresponding point in a target frame which in general will not be on a pixel center. A trajectory will continue the motion from the reference frame across a sequence of frames. That is, the reference frame may be the only time the trajectory is at a pixel center.

In some cases trajectories exist in a low rank subspace. The cases enumerated by Irani [12] include affine motion, weak perspective and orthographic camera models. Generally speaking, anything without a strong perspective change i.e. large depth

© The Author(s) 2016
41
J. Gibson and O. Marques, *Optical Flow and Trajectory Estimation Methods*,
SpringerBriefs in Computer Science, DOI 10.1007/978-3-319-44941-8_4

change relative to overall depth of the scene would stay in a low rank. However, strong perspective changes in at least part of an image sequence are quite ordinary. Conversely, there is generally a significant amount of content which does live in a low rank subspace such as the background objects in a scene.

In this chapter we demonstrate a new method of computing trajectories which can leverage low rank content where it exists but does not require restricted types of motion or camera models. We call this *Robust Low Rank* (RLR).

Computing a low rank constraint on an image sequence would ordinarily be computationally intractable. We borrow from new work by Liu et al. [13] which uses ideas from compressed sensing to find the low rank decomposition of a matrix from a random submatrix.

The remaining color constancy, spatial, smoothness, and robust linkage are possible because of Condat's work [8] presenting a primal-dual solution for a sum of non-smooth functions. The combination of reduced complexity for the low rank and the parallel optimization algorithm on a GPU, make this solution possible.

In this work we show a novel robust low rank constraint of linear complexity for variational trajectory calculation. We demonstrate how the proposed method achieves quantitative and qualitative trajectory improvements on general image sequences over two frame optical flow and state-of-the-art variational trajectory methods. Additionally, we show how optical flow estimation can be improved by using multiple frames and a robust low rank constraint.

4.2 Previous Work

Several previous efforts at multiframe optical flow [14, 15, 23, 26] have tried to extend spatial smoothing into spatio-temporal smoothing. While this works for very small motion, in general it fails because motion is not constrained to a Nyquist sampling rate in the way it is spatially, resulting in local temporal derivatives that do not correspond to the actual motion.

Multiple attempts reported in the literature [3, 4, 16, 21] have used some type of parameterization for optical flow and trajectories. These flow models are generally restrictive and are not helpful with general motion.

Recently there has been new research using Irani's [12] observations on low rank trajectories. On one end of the spectrum is Ricco and Tomasi [18, 19] which use a sophisticated method to find a low rank basis while masking occluded portions from each frame. Some of their results require hours of computation per frame. On the opposite end, Garg et al. 's *Multi-Frame Subspace Flow* (MFSF) makes no attempt at finding occlusions and shows a GPU-based result that runs in near real time.

Ontologically, our method is most related to Garg et al. in that it is a variational method to find a low rank trajectory with a TV-L^1 color constancy term. This is in contrast to [19, 20] which tease apart the trajectories to find ones that stop and match restarts after. In Garg et al. and ours, similar to the TV-L^1 treatment of occlusions in two-frame optical flow, rather than identifying the occluded area and masking it away, a robust matching is used everywhere. However, the similarity between our work and Garg et al. 's ends there. They require the trajectory be constructed from R terms of a DCT or precomputed PCA basis. Our method finds a low rank trajectory concurrent with matching the color constancy and smoothness constraints.

Garg et al. [9] enforces a low rank trajectory to sequences which fit an orthographic camera model. They consider a *hard constraint* and a *soft constraint*. The hard constraint requires the trajectories $U = PQ$. They demonstrate better results with their soft constraint which allows a small difference $||U - PQ||_2$ between the strictly low rank trajectory and the actual computed trajectory. This non-exact low rank trajectory still requires a close fit to low rank model because of the coefficients used and the intolerance of the ℓ_2 norm to outliers. We show that by using an ℓ_1 norm instead we are robust to trajectories that may have an underlying low rank base but have still significant deviations in parts.

These previous efforts use either a predefined DCT or PCA basis to compute a low rank representation of the trajectories. Limiting the number of basis vectors employed sets the rank of the result. Since this low rank representation is tightly coupled to the trajectory produced, the rank of the trajectory must be known.

The low rank constraint work of Cai et al. [7] has been successfully applied to matrix completion problems such as the Netflix challenge. At the foundation of these methods is an SVD thresholding. In the application of low trajectories the matrix is on the order of $2F \times N$ where F is the number of frames and N is the number of pixels per image. The complexity of this SVD $\mathcal{O}(F^2 N)$ becomes quickly intractable to embed in an iterative optimization algorithm.

Wright et al. [24] introduced a *Robust Principal Component Analysis* (RPCA) for matrices where M is the observed matrix which has been corrupted but has an underlying low rank structure where $M = L + S$ where L is low rank and S is sparse. L and S are unknown. The RPCA algorithm has found some applications in imaging, few in video because it becomes computationally intractable as M increases in size. This is generally because of an SVD operation that is repeated during each iteration of the optimization.

A typical approach to the RPCA problem uses the *Alternating Direction Method* (ADM) [25]. This is an efficient method that uses the well-known fact that the nuclear norm is the tightest convex relaxation of matrix rank. This is similar to the ℓ_1 norm being the convex relaxation of sparsity. While efficient, its complexity is still limited typically by the SVD or equivalent.

Recent work by Liu et al. [13] seeks to solve the RPCA problem of recovering a low rank matrix L from a matrix M where $M = L + S$ and S is known to be sparse. They randomly extract a small matrix subsampled from the original. Applying a compressed sensing theorem, they show that solving the RPCA problem on the small matrix gives the same low rank as the full size matrix with overwhelming probability. This is true as long as the randomly selected submatrix is adequately oversampled relative to the rank. They then use the sparsity assumption to deduce the low rank coefficients for the remaining rows and columns.

While the SVD is the typical limiting agent in a RPCA algorithm, Liu et al. point out that while there are other methods that avoid the SVD they are still unable to reduce the complexity due to matrix-matrix multiplies for example.

4.3   Our Work

Irani [12] proved that trajectories are of low rank in a variety of cases such as: orthographic, affine, or weak-perspective camera models, when the field of view and depth variations are small, and when an *instantaneous motion model* is used. This low rank property, particularly of the orthographic camera model, has been used recently by Ricco and Tomasi [19] and Garg et al. [9] to improve trajectory computation. With general perspective camera and motion, the low rank constraint no longer holds. It is ordinary that the trajectory in some part of an image sequence will not be low rank. Forcing a low rank constraint in this instance worsens the calculated trajectory.

Our algorithm uses a low rank constraint that is coupled to the flow estimation via the robust ℓ_1 norm. This tries to find a low rank solution but if one does not exist then it tolerates deviations in a robust manner.

Extending our patch-based flow work [10] to trajectories we minimize:

$$E_{data} + \lambda E_{regSp} + \alpha E_{lowRank} + \tau E_{link} \tag{4.1}$$

where

$$E_{data} = \int_{\Omega} \sum_{n=1}^{F} |I(x + u(x, n), n) - I(x, n_0)|_1 \, dx \tag{4.2}$$

$$E_{regSp} = \int_{\Omega} \sum_{i=1}^{F} g(x)|\nabla u(x, n)|_{2,1} \, dx \tag{4.3}$$

$$E_{lowRank} = \int_{\Omega} \text{rank}(L(x)) \, dx \tag{4.4}$$

$$E_{link} = \int_{\Omega} |U(x) - L(x)|_1 \, dx \tag{4.5}$$

E_{data} is the color constancy term. E_{regSp} is the isometric total variation penalty which is reduced at image edges by $g(x) = \exp(-c_g|I(x, n_0)|^2)$. E_{link} is the robust linkage term between the color constancy and spatially regularized value U and the low rank approximation L. The objective function, Eq. 4.1 is non-convex because of the color constancy and low rank terms. We use a linearized version of the color constancy and the nuclear norm ($| \cdot |_*$) as a convex approximation to the rank.

We break this optimization into two subproblems and iterate between them. First, we fix L then solve for U.

$$\min_{\mathcal{U}}\{E_{data} + \lambda E_{regSp} + \tau E_{link}\} \tag{4.6}$$

Secondly, we fix U then solve for L.

$$\min_{L}\{\alpha E_{lowRank} + \tau E_{link}\} \tag{4.7}$$

These two minimization steps are discussed in more detail in [11].

4.4 Experimental Results

This section contains quantitative and qualitative results that demonstrate improvement over two-frame optical flow and a state-of-the-art trajectory calculation method.

The lack of benchmarks for trajectories makes quantitative comparisons difficult. To apply a meaningful low rank constraint one needs a clip length much greater than the rank constraint. Middlebury has a maximum of eight frames with a single optical flow ground truth frame. The MPI-Sintel training sequences have up to 50 frames which is of adequate length and has optical flow ground truth for each frame to frame transition but does not have trajectories.

The MPI-Sintel benchmarks [1, 6] use difficult, realistic computer generated graphics sequences. Since they are animated sequences, it was possible for the authors to generate exact optical flow ground truth. It is noteworthy that many of the top ranked optical flow algorithms perform poorly on these sequences. We used the "final" version of the rendered training set for our tests. This version is the most realistic and most difficult with specular reflections, camera depth of field blur, and atmospheric conditions.

In the absence of trajectory ground truth for the MPI-Sintel sequences, we match the trajectory to optical flow at the only place they are identical, that is, immediately following the reference frame. We evaluate the Average Endpoint Error (Avg EE) [2] of our generated trajectories and compare against other methods. Since a measure of the trajectory at this one point does not guarantee anything about the remainder of the sequence, we show qualitative results later using additional frames.

We choose the center-most frame of a clip as a reference but still most of the MPI-Sintel sequences are not ideal for trajectory calculation because fast moving content in the reference frame is no longer present in the target frames. Trying to claim a trajectory in these instances is flawed, but at a minimum we show improvement over the simple two frame optical flow calculation.

Some of the sequences were not used because they did badly (Avg EE > 10) with all of the methods tried. Some of the clips had large motion outside of the TV-L^1 grasp. In some of the sequences, the content in the reference frame is largely absent in the remainder of the clip. A trajectory has a full rank of $2F$ where F is the number of frames. Trajectories within the 5–10 frames range are too short to benefit from improvements due to the enforcement of a low rank constraint.

Table 4.1 Avg EE with occlusion mask of MPI-Sintel sequences comparing our RLR results with a 2-Frame TV-L^1, and Garg et al.'s MFSF

Sequence	RLR	2-Frame	MFSF
Alley_1	2.4626	**2.0940**	3.1120
Alley_2	**0.4354**	0.7463	3.2124
Ambush_5	4.5163	**3.4874**	8.5627
Ambush_7	0.3040	**0.2914**	1.5561
Bamboo_1	**0.3876**	0.4295	1.7742
Bamboo_2	**0.4767**	0.5035	0.8262
Bandage_1	**0.4564**	0.7336	2.6576
Bandage_2	**0.4686**	0.5538	2.1927
Mountain_1	**1.7496**	2.7379	3.6332
Shaman_2	**0.2958**	0.2987	1.4862
Shaman_3	**0.3272**	0.6553	0.5984
Sleeping_1	**0.1581**	0.2910	0.8680
Sleeping_2	**0.1100**	0.1435	0.7484

The best results for each sequence are shown in **bold**

Table 4.2 RMS end point error, of flag sequence developed by Garg et al.

Method	Orig.	Occlusions	Gauss. Noise	S&P Noise
RLR	2.39	2.55	5.32	4.84
MPSF$_{I_{2F}}$	1.13	1.43	1.83	1.60
ITV-L^1	1.43	1.89	2.61	2.34
LDOF	1.71	2.01	4.35	5.05
NR Reg	1.24	1.27	1.94	1.79

Other results are from [9]. MFSF$_{I_{2F}}$ is essentially a 2-frame degenerate version of Garg et al., ITV-L^1 by Wedel et al. [22], LDOF by Brox and Malik [5], Non-Rigid Registration by Pizarro and Bartoli [17]

Table 4.1 shows our results on MPI-Sintel sequences, which are superior to the two frame optical flow calculation in almost every case. This is consistent with what one would expect, that looking forward and back should enable a better prediction on the next frame. We are, however, looking across all the (50 usually) frames to make this prediction.

There is a general lack of trajectory benchmarks with ground truth. There is a trail of *ad hoc* quantitative methods including PSNR of warped images, average length of trajectories [19], concatenation of the reverse sequence [20] to produce a zero vector ground truth. These authors all acknowledge their weak metrics, as we have ours. Garg et al. created and published a trajectory ground truth from a synthetic flag sequence. While it may not be ideal, it seemed negligent to ignore it.

In Table 4.2 we show result of our RLR method compared to Garg et al. 's gray scale results on the synthetic flag sequence. It is worth noting that the only gray scale method result that Garg et al. published was the MPSF$_{I_{2F}}$ which degenerates to a simple two-frame flow method. We would assume that they did not find an

sleeping_1, ref frame = 25, target frame = 15   alley_2, ref frame = 25, target frame = 19

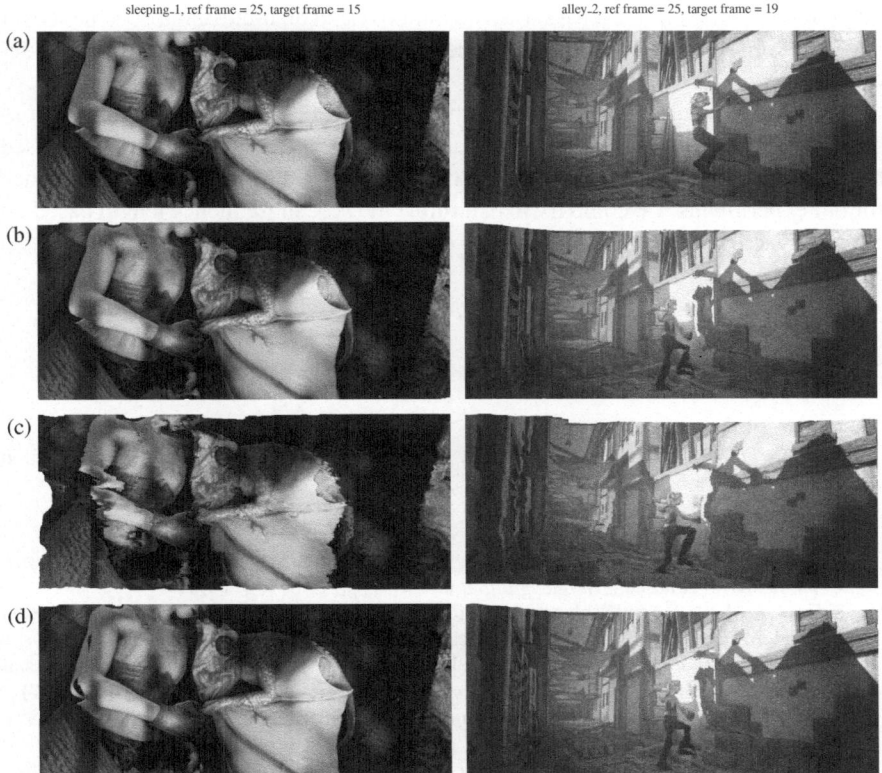

Fig. 4.1 Two sequences from MPI-Sintel [1]. The first row **a** is the reference frame, the subsequent rows **b–d** is the target frame warped back to the reference frame with **b** RLR, **c** MFSF (Garg et al.), **d** TV-L^1. From [11]. Reproduced with permission from Springer

improvement with a low rank constraint for gray scale. We perform generally a little worse than the others but on the same order of magnitude of these state-of-the-art methods. Garg et al. showed low rank improvements by changing to a color vector based algorithm. This enhancement is expected to also improve our results.

In Fig. 4.1 we show two examples of our qualitative improvements using the MPI-Sintel sequences. We compare the fidelity of warped images in rows (b-d) to the reference frame in in row (a). In the first column, our results (b) are clearly best (notice the tears in the arm of row (d)). In the second column, while all methods mistake the fast-moving girl, our RLR approach better preserves the structure of the left wall than either of the other methods.

4.5 Concluding Remarks

We have presented in this chapter a new method to compute trajectories and improve optical flow using multiple frames. We utilize random sampling to reduce the large data complexity. Our choice of convex non-smooth optimization allows a parallel

GPU implementation. We have shown quantitative and qualitative improvement over TV-L^1 baselines for MPI-Sintel sequences as well as competitive results with state-of-the-art trajectory methods.

There are two possible extensions to the current work. A short-term natural extension of this work would be to incorporate the color vector improvement proposed by Garg et al. [9]. In the longer term, rather than using all frames in each sequence for our experiments, we could try to determine the best subsequence length based on content shared across the image sequence.

References

1. MPI Sintel Flow Dataset. http://sintel.is.tue.mpg.de/, 2014
2. S. Baker, D. Scharstein, J.P. Lewis, S. Roth, M.J. Black, R. Szeliski, A database and evaluation methodology for optical flow. Int. J. Comput. Vis. **92**(1), 1–31 (2011)
3. M. Black, Recursive non-linear estimation of discontinuous flow fields, in *Computer Vision—ECCV'94*, pp. 138–145 (1994)
4. M.J. Black, P. Anandan, Robust dynamic motion estimation over time, in *Proceedings of Computer Vision and Pattern Recognition, CVPR-91*, pp. 296–302 (1991)
5. T. Brox, J. Malik, Large Displacement optical flow: descriptor matching in variational motion estimation. IEEE Trans. Pattern Anal. Mach. Intell. **33**(3), 500–513 (2011)
6. D.J. Butler, J. Wulff, G.B. Stanley, M.J. Black, A naturalistic open source movie for optical flow evaluation, in *European Conference on Computer Vision (ECCV)*, pp. 611–625 (2012)
7. J. Cai, E. Candès, Z. Shen, A singular value thresholding algorithm for matrix completion. SIAM J. Optim. pp. 1–26 (2010)
8. L. Condat, A generic proximal algorithm for convex optimization—application to total variation minimization. Signal Process. Lett. IEEE **21**(8), 985–989 (2014)
9. R. Garg, A. Roussos, L. Agapito, A variational approach to video registration with subspace constraints. Int. J. Comput. Vis. **104**, 286–314 (2013)
10. J. Gibson, O. Marques, Sparse regularization of TV-L1 optical flow, in *ICISP 2014, vol. LNCS, 8509 of Image and Signal Processing*, ed. by A. Elmoataz, O. Lezoray, F. Nouboud, D. Mammass (Springer, Cherbourg, France, 2014), pp. 460–467
11. J. Gibson, O. Marques, Sparsity in optical flow and trajectories. Image Video Process. Signal, 1–8 (2015)
12. M. Irani, Multi-frame correspondence estimation using subspace constraints. Int. J. Comput. Vis. **48**(153), 173–194 (2002)
13. R. Liu, Z. Lin, Z. Su, J. Gao, Linear time principal component pursuit and its extensions using ℓ_1 filtering. Neurocomputing **142**, 529–541 (2014)
14. D. Murray, B.F. Buxton, Scene segmentation from visual motion using global optimization. IEEE Trans. Pattern Anal. Mach. Intell. PAMI **9**(March), 220–228 (1987)
15. H.H. Nagel. Extending the 'oriented smoothness constraint' into the temporal domain and the estimation of derivatives of optical flow, in *Proceedings of the First European Conference on Computer Vision* (Springer, New York, Inc., 1990), pp. 139–148
16. T. Nir, A.M. Bruckstein, R. Kimmel, Over-parameterized variational optical flow. Int. J. Comput. Vis. **76**(2), 205–216 (2007)
17. D. Pizarro, A. Bartoli, Feature-based deformable surface detection with self-occlusion reasoning. Int. J. Comput. Vis. **97**(1), 54–70 (2011)
18. S. Ricco, C. Tomasi. Dense lagrangian motion estimation with occlusions, in *2012 IEEE Conference on Computer Vision and Pattern Recognition*, pp. 1800–1807 (2012)
19. S. Ricco, C. Tomasi, Video motion for every visible point, in *International Conference on Computer Vision (ICCV)*, number i (2013)

20. P. Sand, S. Teller, Particle video: long-range motion estimation using point trajectories. Int. J. Comput. Vis. **80**(1), 72–91 (2008)
21. S. Volz, A. Bruhn, L. Valgaerts, H. Zimmer, Modeling temporal coherence for optical flow, in *2011 International Conference on Computer Vision (ICCV)*, pp. 1116–1123 (2011)
22. A. Wedel, T. Pock, C. Zach, H. Bischof, D. Cremers, An improved algorithm for TV-L 1 optical flow, in *Statistical and Geometrical Approaches to Visual Motion Analysis*, pp. 23–45 (Springer Berlin, Heidelberg, 2009)
23. J. Weickert, C. Schnörr, Variational optic flow computation with a spatio-temporal smoothness constraint. J. Math. Imaging Vis. 245–255 (2001)
24. J. Wright, P. Yigang, Y. Ma, A. Ganesh, S. Rao, Robust principal component analysis: exact recovery of corrupted low-rank matrices via convex optimization, in *NIPS*, pp. 1–9 (2009)
25. X. Yuan, J. Yang, Sparse and low-rank matrix decomposition via alternating direction methods. Optimization (Online), 1–11 (2009)
26. H. Zimmer, A. Bruhn, J. Weickert, Optic flow in harmony. Int. J. Comput. Vis. **93**(3), 368–388 (2011)